Cambridge Elements

Elements of Paleontology
Series Editor-in-Chief
Brenda R. Hunda
Cincinnati Museum Center

CENOZOIC PALEOBOTANICAL RESOURCE INVENTORY OF THE NATIONAL PARK SYSTEM

T. P. Matel, *University of Michigan*

I. B. Huegele, *Idaho State University*

C. R. Cace, *University of Texas at Austin*

K. M. M. Bober, *Texas A&M University*

L. D. Boucher, *University of Texas at Austin*

V. E. McCoy, *University of Wisconsin–Milwaukee*

E. J. Hermsen, *Milwaukee Public Museum*

S. R. Manchester, *University of Florida*

C. C. Visaggi, *Georgia State University*

J. S. Tweet, *National Park Service*

V. L. Santucci, *National Park Service*

Shaftesbury Road, Cambridge CB2 8EA, United Kingdom

One Liberty Plaza, 20th Floor, New York, NY 10006, USA

477 Williamstown Road, Port Melbourne, VIC 3207, Australia

314–321, 3rd Floor, Plot 3, Splendor Forum, Jasola District Centre,
New Delhi – 110025, India

Cambridge University Press is part of Cambridge University Press & Assessment,
a department of the University of Cambridge.

We share the University's mission to contribute to society through the pursuit of
education, learning and research at the highest international levels of excellence.

www.cambridge.org
Information on this title: www.cambridge.org/9781009770484

DOI: 10.1017/9781009770477

© T. P. Matel, I. B. Huegele, C. R. Cace, K. M. M. Bober, L. D. Boucher, V. E. McCoy,
E. J. Hermsen, S. R. Manchester, C. C. Visaggi, J. S. Tweet, and V. L. Santucci 2026

This publication is in copyright. Subject to statutory exception and to the provisions
of relevant collective licensing agreements, no reproduction of any part may take
place without the written permission of Cambridge University Press & Assessment.

When citing this work, please include a reference to the DOI 10.1017/9781009770477

First published 2026

A catalogue record for this publication is available from the British Library

ISBN 978-1-009-77048-4 Hardback
ISBN 978-1-009-77050-7 Paperback
ISSN 2517-780X (online)
ISSN 2517-7796 (print)

Additional resources for this publication at www.cambridge.org/matel-et-al

Cambridge University Press & Assessment has no responsibility for the persistence
or accuracy of URLs for external or third-party internet websites referred to in this
publication and does not guarantee that any content on such websites is, or will remain,
accurate or appropriate.

For EU product safety concerns, contact us at Calle de José Abascal, 56, 1°, 28003
Madrid, Spain, or email eugpsr@cambridge.org

Cenozoic Paleobotanical Resource Inventory of the National Park System

Elements of Paleontology

DOI: 10.1017/9781009770477
First published online: February 2026

T. P. Matel, *University of Michigan*
I. B. Huegele, *Idaho State University*
C. R. Cace, *University of Texas at Austin*
K. M. M. Bober, *Texas A&M University*
L. D. Boucher, *University of Texas at Austin*
V. E. McCoy, *University of Wisconsin–Milwaukee*
E. J. Hermsen, *Milwaukee Public Museum*
S. R. Manchester, *University of Florida*
C. C. Visaggi, *Georgia State University*
J. S. Tweet, *National Park Service*
V. L. Santucci, *National Park Service*

Author for correspondence: T. P. Matel, tmatel@umich.edu

Abstract: The National Park System (NPS) encompasses geological exposures that preserve globally significant paleobotanical resources. These paleobotanical resources represent a broad temporal, geographic, stratigraphic, and taxonomic distribution, and pose a variety of management, research, and curation concerns. In this Element, the authors present a baseline inventory of the Cenozoic paleobotany of the NPS as a first step in stimulating new research, curation, and outreach projects that utilize these resources. The authors describe the stratigraphic, taxonomic, spatial, and temporal distribution of Cenozoic paleobotanical resources in 74 National Park units, and show that these resources vary widely in their significance and management needs. Their baseline inventory elucidates what resources need intensified management protocols and celebrates the success stories of NPS paleontological resource management that make NPS lands an essential archive of North American paleobotanical history.

Keywords: paleobotany, national parks, North America, paleoecology, plant evolution

© T. P. Matel, I. B. Huegele, C. R. Cace, K. M. M. Bober, L. D. Boucher, V. E. McCoy, E. J. Hermsen, S. R. Manchester, C. C. Visaggi, J. S. Tweet, and V. L. Santucci 2026

ISBNs: 9781009770484 (HB), 9781009770507 (PB), 9781009770477 (OC)
ISSNs: 2517-780X (online), 2517-7796 (print)

Contents

1	Introduction	1
2	Scope of the Review	7
3	Park Inventories	8
4	Conclusion	85
	References	88

Supplementary Tables are available to access online at www.cambridge.org/matel-et-al.

1 Introduction

Fossils are a rich source of information about the history of the Earth and can be used to test hypotheses about mass extinctions, biogeography, the diversification of clades and communities, the effects of climate change on evolution and extinction, and the effects of evolution and extinction on climate. Fossils are also a finite natural resource, an important aspect of a nation's natural heritage, and a didactic medium that engages public audiences in discourse about the history of life on Earth. The fossil record of the National Park System (NPS) exists precisely at the intersection between fossils as a body of scientific information and fossils as public natural heritage, and the National Park Service balances the multidimensional significance of federal fossil resources through its four core management activities: inventory, monitoring, research, and curation.

Here, we present a comprehensive inventory of Cenozoic paleobotanical resources that details the spatial, stratigraphic, taxonomic, and temporal scope and significance of fossil plants preserved in 74 of the 288 National Park Service areas with paleontological resources. This inventory is the first installment of a complete inventory of Cenozoic, Mesozoic, and Paleozoic paleobotany of the NPS, and the product of a collaborative effort among four student–mentor pairs and NPS scientists sponsored by the Paleontology in the Parks Fellowship Program of the National Park Service and Paleontology Society.

1.1 Cenozoic Paleobotany in North America

The Cenozoic paleobotanical record of North America preserves evidence of the modernization of plant families and terrestrial ecosystems, extinct plant lineages and plant communities without modern ecological analogs, and the response of plants to global events such as the Cretaceous–Paleogene (K–Pg) mass extinction, Paleocene–Eocene Thermal Maximum (PETM), and climatic optima. This history has been written in part based on fossil plants now under the stewardship of the NPS, and provides a rich context for interpreting the significance of new or understudied NPS fossil plant resources. In this section, we summarize key patterns present in the Cenozoic paleobotanical record of the USA in relation to four themes with particular relevance for interpreting NPS resources: recovery from mass extinction, angiosperm diversification, climate change, and intercontinental dispersal. In doing so, we highlight the paleobotanical resources of NPS units that have contributed or have the potential to contribute to a greater understanding of the evolutionary, climatic, and physiographic history of North America.

1.1.1 Extinction Recovery

The collision of a bolide asteroid with the Yucatan Peninsula about 66 Ma prompted a restructuring of terrestrial ecosystems through instantaneous and widespread extinction of species followed by heterogeneous pathways of recovery (Alvarez et al., 1980; McElwain and Punyasena, 2007; Nichols and Johnson, 2008; Wilf et al., 2023). Mass extinction among plants at the K–Pg boundary has been documented from latest Cretaceous and earliest Paleocene outcrops in the Williston Basin in North Dakota (Wilf and Johnson, 2004), the Denver Basin in Colorado (Barclay et al., 2003; Lyson et al., 2019), and the Raton and San Juan basins in southern Colorado and northern New Mexico (Tschudy et al., 1984; Wolfe and Upchurch, 1986, 1987; Flynn and Peppe, 2019). Palynological sequences across the K–Pg boundary indicate a spike in fern spore abundance and paucity of angiosperm pollen immediately above the iridium anomaly that marks the boundary and extinction of 20–30% of pollen taxa (Tschudy et al., 1984; Johnson et al., 1989; Barclay et al., 2003). Macrofloral assemblages offer greater taxonomic resolution than palynofloras, and comparison of latest Cretaceous and earliest Paleocene leaf assemblages indicates even higher plant extinction rates, around 50% (Wolfe and Upchurch, 1986; Wilf and Johnson, 2004; Lyson et al., 2019).

The North American forests that emerged in the first ca. 5 Ma ensuing the K–Pg mass extinction differed markedly in physiognomic and floristic attributes from the latest Cretaceous forests that they replaced, and showed geographic heterogeneity among themselves. Macrofossil evidence from the Raton and Williston basins suggests an increased importance of deciduous trees in Early Paleocene mid-latitude floras based on community-level shifts in leaf functional traits (Wolfe and Upchurch, 1987; Blonder et al., 2014), and this inference meshes well with the prominence of families such as Betulaceae, Cornaceae, Juglandaceae, and Platanaceae, which are common in the temperate deciduous forests of the Northern Hemisphere today (Manchester, 2014). These lineages, alongside taxodiaceous conifers, Cercidiphyllaceae, and Hammamelidaceae, were the dominant components of backswamp and alluvial ridge habitats during the first several million years following the bolide impact (Wing, 1998), and have been interpreted to comprise a low-diversity and spatially homogenous flora spanning from southern Alberta to central Colorado (Johnson and Hickey, 1990; McIver and Basinger, 1993; Wing, 1998; Barclay et al., 2003; Wilf and Johnson, 2004; Manchester, 2014). However, recent reports from earliest Paleocene strata in the San Juan Basin and the Corral Bluffs site in the Denver Basin have documented higher diversity floras within the first million years of the Paleocene (Flynn and Peppe, 2019;

Lyson et al., 2019). These floras comprise a distinct set of taxa, including the earliest legume fossil, and have physiognomic attributes comparable to modern seasonal tropical forests.

Yukon–Charley Rivers National Preserve and Denali National Park and Preserve in central Alaska, and Big Bend National Park in southern Texas each include fossiliferous sequences of Upper Cretaceous and Lower Paleocene rocks (see Section 3 for further details). Increased monitoring, reconnaissance, and research in these units has the potential to further increase our understanding of the geographic heterogeneity in species' responses to the extinction event and subsequent ecological recovery.

1.1.2 Angiosperm Diversification

The Cenozoic is a time interval characterized by diversification within extant families of flowering plants (Ramírez-Barahona et al., 2020; Zuntini et al., 2024) and their rise to ecological dominance (Wing and Boucher, 1998; Carvalho et al., 2021). The Cenozoic strata of the USA contribute to our understanding of this diversification through documenting extinct genera with known (e.g., *Florissantia*; Manchester, 1992) or unknown (e.g., *Palibinia*; Manchester et al., 2024) familial affinities, and early occurrences of extant genera that persist (e.g., *Alnus*; Forest et al., 2005) or have gone extinct (e.g., *Pterocarya*; Stults et al., 2022) from the continent. Collectively, these occurrences illustrate that North American vegetation has been deeply affected by extinction, and that the extant flora is composed of families and genera with heterogeneous ages of origin and evolutionary histories.

Fossil plants provide insight into extinction by preserving combinations of morphological characters that cannot be observed today and by occurring beyond the geographic range of their nearest living relatives. Extinct genera with morphologies consistent with living angiosperm families but not matching any extant genus within them formed a prominent component of Paleogene North American floras and cooccurred with or in some cases predated the first occurrences of confamilial extant genera. *Paleobowditchia*, an extinct genus in the Fabaceae subfamily Papilionoideae, whose type locality is in the Lamar River Formation at Yellowstone National Park (Herendeen et al., 2022), attests to the latter pattern, and *Cedrelospermum*, an extinct genus of the Ulmaceae that cooccurs with *Ulmus* in the fossiliferous shales of Florissant Fossil Beds National Monument (Manchester, 1989a), attests to the former. Several extant families appeared and disappeared from the North American flora over the course of the Cenozoic, and others lost phylogenetic diversity (number of higher taxa below family rank) through the extinction of genera that persist outside of North America today. The Cercidiphyllaceae and Trochodendraceae, monogeneric families now restricted to eastern Asia, are represented by extant and extinct genera in the North American fossil record that grew as common components of Paleocene–Miocene floras of the Great Plains, Rocky Mountains, and Pacific Northwest (Crane et al., 1991; Manchester et al., 1991), including those preserved in Paleocene strata of Theodore Roosevelt National Park in North Dakota (Fastovsky and McSweeney, 1991), and Eocene strata of Glacier National Park in Montana (Smith, 2023). The Juglandaceae, represented in North America today by *Juglans* and *Carya*, were once represented by three additional extant genera that today occur only in Eurasia (Manchester, 1989b), as well as numerous extinct genera (Manchester, 1989b, 1991). Thus, despite its ecological importance in the extant North American flora, the Juglandaceae represent a prominent example of a lineage that lost phylogenetic diversity in North America over the course of the Cenozoic.

Extinction did not affect all lineages equally over the course of the Cenozoic, and the US fossil record also illustrates the capacity of plant families and genera to persist and diversify over multimillion-year timescales. *Acer* and *Quercus* are diverse and ecologically dominant genera today, with evolutionary roots deep in the Cenozoic: Occurrences in the Late Paleocene–Early Eocene Copper Lake Formation in Katmai National Park and Preserve, Alaska (*Acer*; Wolfe and Tanai, 1987) and the Middle Eocene (44 Ma) Clarno Nut Beds of John Day Fossil Beds National Monument (*Quercus*; Manchester, 1994) are among the earliest records of each genus. At the family level, fossil histories of the Betulaceae and Rosaceae indicate a net accumulation of higher-level taxa throughout the Cenozoic, a marked contrast to the trajectory of phylogenetic diversity in the Juglandaceae. The Rosaceae today are among the most diverse families in North America (ca. 700 species), and appear to have first established this diversity during the Eocene (DeVore and Pigg, 2007). Latest Early Eocene floras from the Okanogan Highlands of eastern Washington and British Columbia include the oldest known *Prunus* and *Oemleria* flowers (Benedict et al., 2011), along with early or first occurrences of other rosaceous taxa (Wolfe and Wehr, 1988), and the Late Eocene flora of Florissant Fossil Beds National Monument includes early occurrences of *Rosa*, *Cercocarpus,* and several additional rosaceous taxa (Manchester, 2001). In the Betulaceae, all five extant North American genera have persisted since the Paleogene (Crane, 1989; Forest et al., 2005), and some (e.g., *Carpinus*, *Corylus*) show a similar pattern of stratigraphic occurrence to the Rosaceae, with first occurrences in the Okanogan Highlands (Pigg et al., 2003; Pigg and DeVore, 2010).

1.1.3 Climate Change

Global climate of the Cenozoic transitioned from a high-CO_2 greenhouse world that lacked polar ice sheets in the Paleocene and Eocene to a low-CO_2 world with southern ice sheets beginning in the late Eocene and northern ice sheets beginning in the Pliocene (Zachos et al., 2008; Westerhold et al., 2020; Cenozoic CO2 Proxy Integration Project (CenCOPIP) Consortium, 2023). Superimposed on this general trend are rapid and transient warming episodes, such as PETM and other Early Eocene hyperthermal events, prolonged warm periods such as Early and Middle Eocene climatic optima and the Middle Miocene Climatic Optimum, and abrupt shifts in mean climate state such as the Eocene–Oligocene Transition (EOT), Middle Miocene Climatic Transition, and glacial–interglacial cycles of the Quaternary. Fossil floras of appropriate stratigraphic position document two principal patterns in the response of plants to this climatic history: (1) rapid or cyclic climate changes are matched by rapid plant migration, leading to large taxonomic turnover but low rates of extinction; and (2) prolonged or abrupt and directional climate changes are associated with long-term shifts in plant community composition mediated by immigration and/or local extinction and speciation.

The response of North American plant communities to climatic perturbations lasting less than a million years is best known from two time intervals: the PETM, a period of ca. 200,000 years beginning at the Paleocene–Eocene boundary (56 Ma) that was characterized by 4–8°C of greenhouse warming (McInerney and Wing, 2011); and the last ca. 800,000 years, characterized by large fluctuations in polar ice volume and atmospheric CO_2 levels (Petit et al., 1999; Lüthi et al., 2008). Floral records from the Bighorn Basin in Wyoming indicate that the predominant response of woody plants to the climate changes of the PETM was through reversible range shifts; however, local extirpations and immigrations occurred as well, giving forth to novel earliest Eocene plant associations (Wing et al., 2005; Wing and Currano, 2013; Korasidis and Wing, 2023). A similar response of eastern North American vegetation to cycles of icesheet expansion and retraction during the Late Pleistocene and Holocene has been documented from pollen cores taken from a large network of glacial lakes, including one within Cape Cod National Seashore in Massachusetts (Winkler, 1985). Collectively, these records illustrate that the formation and dissolution of plant associations reflect individualistic interactions between species and climatic factors (Jackson and Overpeck, 2000).

In contrast, plant responses to climatic changes occurring over multimillion-year timescales are characterized by widespread extinction and directional shifts in floristic composition and/or vegetation structure. This pattern is well-documented in the Eocene floras (Leopold and MacGinitie, 1972; Wing, 1987) of the Rocky Mountain Region that record a progressive loss of Late Paleocene and Early Eocene taxa, many with East Asian biogeographic affinities (e.g., *Metasequoia*; Wing, 1987) and influx of taxa with neotropical biogeographic affinities (e.g., Myrtaceae; Manchester et al., 1998). This shift in floristic composition has been correlated with regional drying in the Rocky Mountain Region and differs from the Eocene record of the Pacific Northwest, which retained a wetter climate and tropical to subtropical "East Asian" lineages up until the Late Eocene–Early Oligocene (Leopold and MacGinitie, 1972; DeVore and Pigg, 2010). The Middle Eocene (44 Ma) Clarno and Early Oligocene (33 Ma) Bridge Creek floras from Oregon, preserved in John Day Fossil Beds National Monument, illustrate major changes in the makeup of the Pacific Northwest flora associated with a gradual trend of climatic cooling and drying in the Middle–Late Eocene, punctuated by the abrupt climatic deterioration of the EOT (Retallack, 1992; Wolfe, 1992). This period of climate change has been interpreted as a major factor in the regional replacement of evergreen angiosperms by deciduous angiosperms and conifers, and the regional extinction of numerous taxa that persist today in the eastern USA and eastern Asia (MacGinitie, 1953; Meyer and Manchester, 1997).

1.1.4 Intercontinental Dispersal

Three land connections influenced the potential for intercontinental dispersal via biogeographic corridors during the Cenozoic in North America: the Bering Land Bridge (BLB) between Alaska and Russia; the North Atlantic Land Bridge (NALB) between eastern North America and Western Europe; and the Isthmus of Panama that connected the Americas. The BLB lay at high northern latitudes, likely between 60 and 80°N, and served as a viable corridor throughout the Cenozoic, despite variability in the degree of continuity and local climate of land across the Bering Strait (Tiffney and Manchester, 2001). Migration across the NALB is hypothesized to have occurred across a northern pathway, Svalbard to Greenland to Ellesmere Island (McKenna, 1975, 1983), and a southern pathway, Scotland to the Faroe Islands to Iceland to Greenland to North America (Tiffney, 1985; Tiffney and Manchester, 2001; Denk et al., 2011). Current tectonic and paleobotanical evidence (reviewed by Denk et al., 2011) suggests that migration via the southern NALB occurred up until the Late Miocene (10–6 Ma). The Isthmus of Panama formed a persistent terrestrial landscape between Central and North America around 3.5 Ma, and was the result of a complex sequence of events, beginning with the formation of a large island in the Late Eocene (Jaramillo, 2018). The geologic history of the land bridge and the frequency of biotic interchange prior to its formation is an active area of debate (Montes et al., 2015; O'Dea et al., 2016).

The inference of migration routes for taxa that now or once occurred on multiple Laurasian (North America, Europe, Asia) continents are based on the geo-stratigraphic distribution of fossil taxa and the present distribution of their living relatives (Manchester, 1999). The fossil record of North America includes immigrant taxa with earliest stratigraphic occurrences in Europe or Asia (e.g., *Betula*), supporting dispersal into North America by the NALB or BLB, as well as migrant taxa with earliest occurrences in North America that subsequently occur in Asia and Europe (e.g., *Cornus*). Among extant lineages, legacies of intercontinental dispersal have contributed to modern Holarctic distributions (e.g., *Acer*), and Asian–North American disjunctions (e.g., *Nyssa*). Differential extinction following range expansion via past dispersal corridors has also produced patterns of paleo-endemism among taxa that were widespread as fossils but only persist on one continent (e.g., *Eucommia* in Asia, *Comptonia* in North America), and "range-reversal" for taxa that persist on a different continent from their earliest fossil occurrence (e.g., *Cyclocarya*).

Early and Middle Eocene floras preserved in several Alaskan national parks provide insight into the vegetation structure near the migration corridor of the BLB and provide the opportunity to catch modern Holarctic taxa in the act of migration. Early Eocene plants from Wrangell–St. Elias National Park and Preserve include lineages that are mainly tropical and evergreen today, such as the Icacinaceae, Lauraceae, and Annonaceae (Wolfe, 1966, 1977), implying that the greenhouse climate of the Early Cenozoic allowed some members of these families to reach high latitudes despite the increased seasonality of insolation. Further systematic and paleoecological study of the floras within Alaskan national parks has the potential to improve our understanding of the origin of modern biogeographic patterns among Northern Hemisphere plants and of ecosystem function within non-analog high-latitude evergreen floras.

2 Scope of the Review

Our inventory focuses on occurrences of plant macrofossils (megafossils) in Cenozoic sedimentary rock formations and unlithified sediments that fall within the boundaries of national parks, historical parks, scenic and heritage trails, monuments, preserves, recreation areas, and other types of units administered by the National Park Service. We define plant macrofossils as macroscopic physical evidence of past plant life (Santucci et al., 2024) and include permineralizations, charcoalifications, and compressions and impressions of leaves, flowers, fruits, seeds, stems, branches, and wood. In addition, we selectively include records of plant microfossils, such as pollen, spores, phytoliths, and dispersed cuticle from cores and outcrops and plant ichnofossils and related sedimentary structures, such as root traces and other rhizoliths. The microfossil and ichnofossil records are here reported in instances where their presence indicates potential for macrofossil discovery or has contributed to paleoenvironmental interpretation of plant macrofossil assemblages. We exclude macrofossil occurrences documented from packrat middens (reviewed by Tweet et al., 2012a) and provide less thorough coverage of the substantial body of literature devoted to the Quaternary pollen record.

To compile the NPS Cenozoic plant fossil inventory, we searched peer-reviewed scientific journals, NPS archives (internal park communications, internal reports, etc.), geologic maps, paleontological inventory reports, USGS internal memos ("E&R reports"), popular science articles, museum collection databases, and other relevant databases such as the Alaska Paleontological Database (www.alaskafossil.org/) and the Florissant Fossil Database (https://flfo-search.colorado.edu/). Potential records were checked against NPS boundaries to determine whether a given occurrence was within or outside a given park unit, and for confirmed fossils, we conducted a secondary literature search to the best of our abilities to find associated geochronologic, stratigraphic, taphonomic, and taxonomic information. We present park-wide summaries for the Quaternary and each epoch of the Paleogene and Neogene in Supplementary Tables 1–6 (Supplementary Tables are available to access online at www.cambridge.org/matel-et-al). Text summaries are presented later, organized geographically by the seven NPS legacy regions, and within regions by the 32 Inventory and Monitoring networks.

3 Park Inventories

3.1 Alaska Region

The Alaska Region includes 16 NPS units divided among four inventory and monitoring networks and encompasses 54 million acres, approximately two-thirds of the land area managed by the NPS (Figure 1). Of these 16 parks, 11 contain confirmed records of Cenozoic plant macrofossils, and Yukon–Charley Rivers National Preserve likely contains undocumented Cenozoic plant macrofossils (Kenworthy and Santucci, 2003; Santucci and Kenworthy, 2008; Elder et al., 2009; Santucci et al., 2011).

Alaskan national parks preserve highly significant paleobotanical resources that provide insight into the evolution of the Alaskan flora over the whole of the Cenozoic (Figure 2) but have especially enriched the paleobotanical understanding of arctic floras during the Eocene and Oligocene (Figure 3). Early Eocene floras from the Kulthieth Formation in Wrangell–St. Elias National Park and Preserve as well as floras from the Kushtaka Formation to the south

Figure 1 Map of the 11 NPS units with confirmed Cenozoic paleobotanical resources in the Alaska Region and one unit (YUCH) with potential resources. Interior green shading illustrates the area enclosed by the perimeter of park boundaries plus a 3 km buffer, and exterior shading colored by inventory and monitoring represents the area enclosed by the perimeter of park boundaries plus a 30 km buffer. Unit acronyms are generally formed from the first four letters of one-word unit names (e.g., Katmai National Park and Preserve = KATM) and the first two letters of the first two words for multiple-word unit names (e.g., Bering Land Bridge National Preserve = BELA).

document "boreotopical" forests, a non-analog vegetation type characterized by rainforest-like leaf physiognomy and the presence of some lineages now restricted to the tropics and subtropics of Asia (Wolfe, 1977, 1994a, 1994b; Wing, 1998; Graham, 1999). Middle and Late Eocene floras are known from Aniakchak National Monument and Preserve and Katmai National Park and Preserve on the Alaskan Peninsula (Hollick, 1936; Wolfe, 1966), as well as from the Katalla and Rex Creek floras of Central Alaska (Wolfe, 1992), and illustrate a step in the modernization of the flora, with an increased presence of the broadleaf deciduous lineages (e.g., Betulaceae) characteristic of modern North American forests. Oligocene and Miocene floras, known from NPS units on the Alaskan Peninsula and Central Alaska as well as a number of sites outside of park boundaries (Wolfe, 1977; Wing, 1998), represent further modernization of the flora by complete loss of broadleaf evergreen lineages and their replacement by broadleaf deciduous plants and conifers, with an apparent peak

Figure 2 Summary of the temporal distribution and significance of Cenozoic paleobotanical resources in the Alaska Region. The significance rating of individual parks represents the aggregate significance of all resources within that park and was scored based on criteria such as holotype specimens, plant fossil abundance and preservation quality, and number of publications referencing primary paleobotanical data from the park.

in the diversity of the broadleaf deciduous component during the Early Miocene (Wolfe and Tanai, 1980).

Numerous additional Alaskan paleofloras are preserved outside of the NPS and add important context for interpreting and managing park resources. Much of the primary documentation of these resources comes from the monographic works of Swiss paleobotanist Oswald Heer and USGS paleobotanists Arthur Hollick and Jack Wolfe (Heer, 1869; Hollick, 1930, 1936; Wolfe, 1966, 1977; Wolfe et al., 1966), and the Tertiary (used here for convenience for pre-Quaternary Cenozoic) record in Alaska has been reviewed several times (Wolfe, 1972; Wing, 1998; Graham, 1999).

3.1.1 Arctic Alaska Inventory and Monitoring Network

Cape Krusenstern National Monument

Cape Krusenstern National Monument contains approximately 650,000 acres of land along the northwestern coast of Alaska (Figure 1). The monument was established in 1978 to protect a series of beach ridges along the southern coast (Lanik et al., 2019). Unidentified fossil wood recovered from these ridges is the only report of paleobotanical material within the park (Mason and Ludwig, 1990).

Figure 3 Stratigraphic chart of Tertiary plant-fossil–bearing formations in the Alaska Region.

Kobuk Valley National Park

Kobuk Valley National Park preserves approximately 1.75 million acres of the Kobuk River Valley and borders Noatak National Preserve to the north (Figure 1). The only report of Cenozoic paleobotanical material within the park comes from Epiguruk Bluffs, a locality on the Kobuk River containing fossil vertebrates, mollusks, and wood. Wood samples used for radiocarbon dating constrain the age of the deposits to 37,000–14,000 years before present (ybp) (Hamilton and Ashley, 1983; Hamilton et al., 1984, 1993; Lanik et al., 2019).

Gates of the Arctic National Park and Preserve

Gates of the Arctic National Park and Preserve (GAAR) lies entirely north of the Arctic Circle and is the nation's second largest national park, encompassing nearly 8.5 million acres (Figure 1). Paleobotanical material collected from GAAR includes plant fragments from unnamed lacustrine sediments on the north side of Noatak Valley just north of Okturak Creek (Forester, 1979) and fossil wood from overbank deposits of the west bank of the Killik River (Repenning, 1974).

Bering Land Bridge National Preserve

Bering Land Bridge National Preserve (BELA) covers 2.8 million acres of tundra on the Seward Peninsula of Alaska (Figure 1). Quaternary paleobotanical resources within BELA have been discovered in eroding permafrost sediments (pingo deposits), tephra deposits of the Devil Mountain maar exposed on banks of thermokarst lakes, and in gravel deposits along the banks of the Noxapaga River (Spackman, 1951; Goetchus and Birks, 2001; Wetterich et al., 2012). Just east of BELA in the vicinity of Lava Camp Mine, is a volcanic lava flow dated to 5.7 ± 0.2 Ma (Late Miocene) that preserved a forested floodplain (Hopkins et al., 1971). The museum collection at BELA holds a fossil leaf and fossil liverwort specimen, but these are not linked to locality or age information (Lanik et al., 2019).

Lava Camp Mine: The Lava Camp Mine locality (USGS 11190) in the Imuruk River Valley just east of the park boundary preserves plant macrofossils in alluvial deposits that were buried under an upper Miocene (5.7 Ma) basaltic lava flow (Hopkins et al., 1971). The macrofossil assemblage represents a floodplain forest composed largely of *Picea* and *Betula*. Its taxonomic composition is similar to approximately contemporaneous deposits from the Cook Inlet Region, and together these floras represent the Clamgulchian (Late Miocene–Pliocene) Alaskan time-stratigraphic stage (Wolfe et al., 1966; Hopkins et al., 1971). Macrofossil types include needle leaves, ovulate cones, and seeds; taxa include *Picea* (three spp.), *Pinus*, *Tsuga*, *Betula*, *Symphoricarpos*, *Vaccinium*, *Carex*, and *Cyperus* (Hopkins et al., 1971). The locality also preserves a diverse insect and arachnid assemblage consisting of at least 83 species (Elias et al., 2006). Parts of the southeast corner of the park contain upper Tertiary (Neogene) volcanic deposits (map unit Qtv; NPS, 2015 [unpublished digital geologic map]), which might preserve similar communities.

Noxopaga River: Fossil wood identified as *Populus* and *Picea* was recovered from a 12 m (40 ft) section of Quaternary gravel deposits exposed on the west bank of the Noxopaga River, between Goose and Bussard creeks (Spackman, 1951).

Pingo Deposits: Wetterich et al. (2012) collected samples of pingo deposits from Whitefish maar, Devil Mountain maar, and North and South Killeak maars in the Cape Espenberg lowlands (within BELA) of the Seward Peninsula and from nine other localities (outside BELA) in Arctic and Central Alaska. The combined age range of these samples spans 46,615–8,986 ybp, and the oldest sediments containing identifiable plant macrofossils were constrained to 32,870 ± 220 ybp. The macrofossil assemblage includes twigs, leaves, seeds, and fruits representing gymnosperms and woody and herbaceous angiosperms. Common plant genera include *Betula*, *Carex*, *Hippuris*, *Potamogeton*, *Potentilla*, and various members of the Ericaceae.

Devil Mountain Lake Tephra: The Devil Mountain Lake tephra buried approximately 2500 km^2 (970 mi^2) of land surface, reaching a maximum thickness of 1 m (3 ft) over an approximately 1200 km^2 (460 mi^2) region (Goetchus and Birks, 2001). The tephra was subsequently capped by loess, preserving an in situ tundra landscape, and the sequence has recently become exposed along the banks of thermokarst lakes. Plant macrofossils and organic material used for radiocarbon dating were sampled from twelve different localities in the Cape Espenberg–Devil Mountain area and produced an age range of 18,260–17,700 radiocarbon ybp (ca. 21,500 calendar ybp). This age range places the deposits within the thermal minimum of the Last Glacial Maximum (28,000–14,000 ybp). The macrofossil assemblages are interpreted to represent a sedge-dominated, herb-rich tundra grassland based on the relative ground cover of moss, graminoids, forbs, and shrubs in point-transect collections and the composition of seeds and fruits isolated from sieved sediment. Representative taxa include *Carex*, *Salix*, *Papaver*, and *Potentilla* (Goetchus and Birks, 2001).

3.1.2 Central Alaska Inventory and Monitoring Network

Yukon–Charley Rivers National Preserve

Yukon–Charley Rivers National Preserve (YUCH) encompasses approximately 2.5 million acres (Figure 1). This area protects about 161 km (100 mi) of the Yukon River and the entire Charley River basin. The park's geology spans nearly 1.3 billion years of Earth's history, from the Precambrian to the Cenozoic (Santucci et al., 2011). The park does not contain any confirmed Cenozoic paleobotanical resources; however, it does contain exposures of Upper Cretaceous and Paleocene rocks (map unit TKs; Figure 4), which have yielded Paleocene plant fossils just outside of the park's southern boundary (Hollick, 1930, 1936; Mertie, 1942) and Cretaceous plant fossils from within the park (Knoll, 1976; Tiffney, 1976; Fiorillo et al., 2014).

Unnamed unit (TKs): Plant macrofossils and palynomorphs are preserved in a continuous Upper Cretaceous to lower Cenozoic depositional sequence mapped as unit TKd (Foster, 1976) or, more commonly, as TKs (Mertie, 1930; Knoll, 1976; Dover and Miyaoka, 1988; Miyaoka, 1990). This unit is laterally

Figure 4 Fossil plants from the Alaska Region. (a) *Paratinomiscium conditionalis,* a leaf belonging to the Menispermaceae recovered from the Yakutat Bay locality in Wrangell–St. Elias National Park and Preserve. Image from Hollick (1936). (b) *Corylus harrimani* Knowlton collected from Katmai National Park and Preserve. Scale bar = 5 cm. (c) *Ptermospermites alaskana* collected from Katmai National Park and Preserve. Scale bar = 5 cm. (d) Holotype of *Acer disputabilis* collected from Katmai National Park and Preserve. Scale bar = 1 cm. (e) *Myrica banksiaefolia* var. *curta* collected from Katmai National Park and Preserve. Scale bar = 1 cm. (f) Specimen of *Picea harrimani* collected from Katmai National Park and Preserve. Scale bar = 1 cm. (g) Holotype specimen of *Aesculus arctica* collected from Katmai National Park and Preserve. Scale bar = 5 cm. (h) Holotype specimen of *Hicoria magnifica* collected from Katmai National Park and Preserve. Scale bar = 5 cm. (i) *Alnus evidens,* recovered from the Kukak Bay locality in Katmai National Park and Preserve. This taxon was formerly placed in the genus *Corylus*. Image from Hollick (1936).

continuous from the southeast corner of the park (near Eagle, Alaska) to the northwest of the park (south of Circle, Alaska). It forms a belt approximately 145 km (90 mi) long and parallel to the course of the Yukon River, and its Paleogene deposits are thickest and most laterally extensive in its southern and eastern regions (Mertie, 1942; Tiffney, 1976). The subdivision of the sequence represented by map unit TKs into smaller stratigraphic units is likely warranted but has been hindered by the unit's large areal extent, variable lithology, and widely spaced fossil localities (Tiffney, 1976).

At least 28 plant macrofossil localities are documented from TKs in the Eagle–Circle district (Mertie, 1942). Many of the early discoveries of fossil localities from this formation occurred in the southeastern region of YUCH and just outside the park's southern boundary (Hollick, 1930, 1936; Mertie, 1942). Among these localities are several Early Cenozoic sites along the banks of Seventymile Creek and one of its tributaries, Bryant Creek, which have yielded well-preserved angiosperm leaf fossils, several of which were described as new species (Hollick, 1936). Geologic maps indicate that there is significant potential to find similar localities within the southeastern region of YUCH, which is approximately 8 km (5 mi) to the north.

Denali National Park and Preserve

Denali National Park and Preserve (DENA), encompasses approximately 6 million acres and protects the large glaciers of the Alaska range as well as a diverse megafauna (Figure 1). The park's geology is complex, and study and interpretation by the NPS, USGS, and academic researchers are ongoing (Csejtey et al., 1986; Tomsich et al., 2010; Santucci et al., 2011). Cenozoic paleobotanical material has been recovered from at least three different formations within DENA: an unnamed Lower Eocene map unit labeled unit Tfv, the Upper Oligocene Healy Creek Formation, and the Upper Miocene to Pliocene Nenana Gravel (Figure 4; Csejtey et al., 1986; Santucci et al., 2011). The Cantwell Formation, once considered to be Paleocene in age, has also yielded plant fossils within DENA; however, recent studies (Tomisch et al., 2010) indicate a Late Cretaceous age for the Cantwell Formation.

Fluviatile and subordinate volcanic rocks (Eocene?), unit Tfv: While developing a geologic map for the Healy Quadrangle, Csejtey et al. (1986) documented two plant fossil localities within DENA that are ascribed to an unnamed map unit labeled Tfv. These localities are in the eastern-central area of the park between Costello Creek and the western fork of the Chulitna River (see Csejtey et al., 1986, for further details). The proposed Eocene age of this unit is based only on its sparse plant fossils and correlation with other Early Cenozoic

Alaskan fossil assemblages. Plant fossils from this unnamed unit were found in a shale interbed within conglomerate and include needles of *Metasequoia occidentalis* and fragments of broad-leaved dicotyledons (Csejtey et al., 1986). This sample was studied by paleobotanist Jack Wolfe, who suggested that the dicotyledonous component represented a plant community younger than the Cantwell Formation, which he then believed to be Paleocene in age (J. A. Wolfe, pers. comm. in Csejtey et al., 1986).

Healy Creek Formation: The Healy Creek Formation is the basal unit of the Usibelli Group (also known as the Tertiary coal-bearing group, or map unit Tcu of Nenana Coalfield), which encompasses five formations that span from the Oligocene to Miocene (Figure 4; Wahrhaftig et al., 1969). A collection of plant fossils assigned to this group was made at the top of the section exposed at Dunkle Coal Mine, in the same east-central region of DENA, which also yielded plant fossils from the unnamed unit (unit Tfv) already discussed (Csejtey et al., 1986). This assemblage includes twelve species described from leaf compressions and impressions. *Cercidiphyllum crenatum* and *Alnus evidens* are notable from a biostratigraphic standpoint and suggest that the assemblage belongs to the Angoonian (Late Oligocene) Alaskan time-stratigraphic stage (Wolfe et al., 1966). Other taxa present in this collection include the fern *Onoclea*, the gymnosperm *Metasequoia* cf. *glyptostroboides*, and broad-leaved angiosperms such as *Betula*, *Salix*, *Populus*, and *Corylus*. Elsewhere in DENA, Bruce Reed collected an assemblage including *Metasequoia* cf. *glyptostoboides*, *Alnus evidens*, "*Carya*" *magnifica*, *Equisetum*, and undetermined species of *Alnus* and *Populus* at USGS localities 11123–11125. The localities are assigned to an unnamed unit composed of conglomerate, sandstone, and siltstone interpreted as Oligocene in age on the basis of its plant fossils (Wolfe, 1968). It is possible that this assemblage was recovered from unmapped exposures of the Healy Creek Formation.

Outside of DENA, in the Nenana Coalfield, the Healy Creek Formation has yielded similar assemblages to those recovered from the park; these assemblages also include *Ulmus* and *Quercus* (Wahrhaftig et al., 1969). In the Nenana Coalfield, the Healy Creek deposits are in conformable contact with other paleobotanically productive formations of the Usibelli Group, and there is potential to discover similar sequences in the northeastern region of DENA where the Usibelli Group is exposed in the Savage River drainage and near Sable Pass.

Nenana Gravel: The Nenana Gravel includes conglomerate, sandstone, claystone, and lignite that accumulated during uplift and erosion of the Alaska Range in the Pliocene. The unit is exposed along portions of the park road in the park's northeast corner (Csejtey et al., 1992; Wilson et al., 1998;

Santucci et al., 2011). Within DENA, the only paleobotanical materials recovered from the gravel to date are a *Picea* cone and carbonized wood in bajada deposits (Santucci et al., 2011). Leopold and Liu (1994) reported carbonized plant fragments and fossil fruits of *Trapa* from probable deposits of the Nenana Gravel northeast of the park near Ferry, Alaska.

Wrangell–St. Elias National Park and Preserve

Wrangell–St. Elias National Park and Preserve (WRST) in southeast Alaska is the largest national park in the USA, encompassing more than 13 million acres (Figure 1). The park was established to conserve a region that includes three major mountain ranges – the Chugach, St. Elias, and Wrangell mountains – numerous glaciers, and wildlife habitat ranging from temperate rainforest to tundra. The park also contains significant Cenozoic paleobotanical resources. In 1905, R. S. Tarr was the first to collect plant fossils within what is now WRST, from outcrops of the Paleocene–Eocene Kulthieth Formation along Esker Stream near Yakutat Bay. The collection from this locality was later studied by USGS paleobotanists Arthur Hollick and Jack Wolfe, yielding type specimens of several species (Figure 4; Hollick, 1936; Wolfe, 1977). Later collecting efforts, also led by the USGS, yielded plant specimens from three additional formations: the Eocene–Miocene Poul Creek Formation, the Miocene Frederika Formation, and the Miocene–Pleistocene Yakataga Formation (Figure 4; Wolfe, 1977; Fremd et al., 2003; Santucci et al., 2011).

Kulthieth Formation: Fossil plants from the Kulthieth Formation were recovered in a calcareous sandstone layer at USGS paleobotanical locality 3879 along Esker Stream, which flows into Yakutat Bay in the southeast corner of the park (Hollick, 1936). The stratigraphic position of this locality within the Kulthieth Formation is uncertain. Wolfe (1977) initially referred the Kulthieth Formation to the Middle Eocene based on mollusk fossils occurring in intertonguing marine rocks (Marincovich and McCoy, 1984), but its planktonic microfossils indicate an Early Eocene age, potentially within the Early Eocene Climatic Optimum (Wolfe, 1994a). The fossil assemblage collected at Esker Stream includes ferns, gymnosperms, and angiosperms represented by leaf compressions and impressions and one fossil endocarp (Figure 4). This locality produced type specimens of several species that belong to angiosperm families that do not occur in Alaska today: *Celastrus comparabilis* Hollick in the Celastraceae, *Paratinomiscium conditionalis* in the Menispermaceae (Figure 4), and *Paleophytocrene elytraeformis* in the Icacinaceae; it also produced the lectotype specimen of an extinct species of wood fern, *Dryopteris alaskana*. Wolfe (1977) considered the fossil flora collected from the Kulthieth Formation to represent

"Paratropical Rainforests" based on its high proportion of large and entire-margined leaves and the presence of the families Annonaceae and Icacinaceae, which mainly occur in the tropics today. Climate reconstructions based on leaf physiognomy of 54 species from the Kulthieth Formation estimate that the assemblage lived under the warmest climate of any Alaskan Paleogene flora, with a mean annual temperature of 19.4°C (66.9°F) (Wolfe, 1994a).

Poul Creek Formation: Fossil plants from the Poul Creek Formation were collected from USGS paleobotanical locality 11185, which occurs near Marvine Glacier in the southeast corner of the park (Wolfe, 1977). Molluscan biostratigraphy of the Poul Creek Formation suggests that its sediments were deposited between the Late Eocene and Early Miocene (Figure 3), and the presence of *Alnus evidens* at USGS 11185, an indicator of the Angoonian (Late Oligocene) floral stage in Alaska, suggests an Oligocene depositional age for the formation's fossil-plant–bearing strata (Wolfe, 1977; Marincovich and McCoy, 1984; Santucci et al., 2011). Elsewhere, the Poul Creek Formation is known as a marine unit, but USGS 11185 occurs along a fault in which no marine fossils were found (Stoneley, 1967; Wolfe, 1977). The plant remains recovered from the Poul Creek Formation are fossil leaves assigned to the angiosperms *Alnus evidens* and *Cercidiphyllum crenatum* and the gymnosperm *Metasequoia glyptostroboides* (Wolfe, 1977). Outside of WRST, a well-preserved permineralized walnut, *Juglans lacunosa*, was illustrated from Poul Creek locality USGS 11182 (figure 48 p, q, in Manchester, 1987).

Frederika Formation: The Frederika Formation occurs in the Wrangell Volcanic Fields, and outcrops extend into the northern and eastern regions of the park (Richter et al., 2006). The valley walls of a westward-flowing tributary of Frederika Creek (Skolai Creek) are the type area of the Frederika Formation and the location of USGS localities 9933, 9935, and 9927, each of which falls within park boundaries (Wolfe et al., 1966; Wolfe and Tanai, 1987). The Frederika Formation preserves paleobotanical material in a variety of depositional contexts, including paleosol sequences, thick coal beds, lacustrine deposits, and silty tuffaceous sandstones (Fremd et al., 2003). The plant fossils recovered from the Frederika Formation in the Skolai Creek area include angiosperms common in temperate deciduous forests, such as *Acer*, *Alnus*, *Betula*, and *Populus*, as well as the conifers *Pinus* and *Metasequoia* (Wolfe, 1966; Santucci et al., 2011). Collectively, this assemblage suggests a Miocene age (Wolfe, 1966). Wolfe (1964) remarked that the Pinaceae are better represented in the Frederika Formation than in contemporaneous floras of Cook Inlet and speculated that the difference in plant composition could be linked to the greater topographic relief and elevation of the Wrangell Mountains during the Miocene.

Yakataga Formation: The Yakataga Formation mainly comprises marine sequences but includes two paleobotanical localities (USGS 11183 and 11184) established in strata that have a similar lithology to the marine beds (Wolfe, 1977). Both localities are in the southeastern area of the park, near Yakutat Bay. The locality USGS 11183 occurs in exposures near Haenke Glacier and USGS 11184 occurs in the Pinnacle Hills area. Floral and faunal biostratigraphy of the Yakataga Formation exposures that produced the plant-fossil–bearing horizons suggest that they were deposited in the Miocene (Wolfe et al., 1966; Wolfe, 1977). Respectively, USGS 11183 and 11184 were referred to the Seldovian (Early to Middle Miocene) and Homerian (Middle to Late Miocene) floral stages based on the presence of *Alnus appsii* and *Carpinus cobbii* (Wolfe, 1970, 1977; Addicott et al., 1978). Fossil leaves of the fern *Osmunda* and the gymnosperm *Metasequoia glyptostroboides* were also recovered from the Yakataga Formation in WRST, and a fossil *Fagus* leaf was recovered from the Yakataga Formation in Glacier Bay National Park and Preserve (see Section 3.1.4) (Wolfe, 1977).

3.1.3 Southwest Alaska Inventory and Monitoring Network

Lake Clark National Park and Preserve

Lake Clark National Park and Preserve (LACL) encompasses 4 million acres of a dynamic natural landscape renowned for its coastline along Cook Inlet, glaciers, glacial lakes, rugged mountain peaks, and two active volcanoes, Mount Iliamna and Mount Redoubt (Figure 1). The park preserves one of the most important Mesozoic marine fossil localities in Alaska at Fossil Point, located on the Cook Inlet coast in Tuxedni Bay (Kenworthy and Santucci, 2003). Younger rocks of the Paleocene–Lower Eocene West Foreland Formation and the Oligocene–Middle Miocene Tyonek Formation also crop out along the Cook Inlet coast and contain plant macrofossils (Figure 3; Hollick, 1936; Wolfe et al., 1966).

West Foreland Formation: Exposures of the West Foreland Formation on the north shore of Chinitna Bay in the Iniskin-Tuxedni Region have produced fossil wood and leaves from locality USGS 3505 (Wolfe et al., 1966; Ruga et al., 2020). The West Foreland Formation is considered to be Late Paleocene–Early Eocene in age based on its plant fossils (Magoon et al., 1976a, citing an oral comm. with Wolfe). Calderwood and Fackler (1972) considered it to be the basal formation of the Kenai Group, a sequence of botanically fossiliferous terrestrial deposits spanning the Upper Paleocene to the Pliocene. The holotype specimen of *Ginkgo reniformis* var. *conformis* Hollick was collected from USGS 3505, and other taxa also found at this locality include *Ginkgo biloba*,

Metasequoia glyptostroboides, and *Taxodium tinajorum* (Hollick, 1936; Wolfe et al., 1966). Fossil tree stumps exposed on the north shore of Chinitna Bay were discovered in 2019. They have not been studied taxonomically (or otherwise) and should be a management priority because they are at high risk of being lost to erosion (Ruga et al., 2020). An additional macrofossil locality of the West Foreland Formation has been found just north of the park boundary near Redoubt Point (Magoon et al., 1976b), and there is potential to find additional localities in the Iniskin-Tuxedni Region within LACL.

Tyonek Formation: USGS paleobotanical localities 9760, 9887, 11344, 11355, and 11367 were established from exposures of the Tyonek Formation within park boundaries at Redoubt Point (Wolfe et al., 1966; Wolfe and Tanai, 1987). Like the Kukak Bay localities in Katmai National Park and Preserve (see below), these deposits are considered to be Upper Oligocene based on the presence of taxa characteristic of the Angoonian floral stage, such as *Alnus evidens* and *Corylus harrimanii* (Wolfe et al., 1966; Wolfe and Tanai, 1980). Other taxa collected from the Redoubt Point flora include the conifers, *Glyptostrobus europaeus* and *Metasequoia glyptostroboides*, and angiosperms, *Vaccinium, Populus* cf. *kenaina, Carpinus cappensis,* and the holotype specimen of *Acer kenaicum*. Wolfe (1994b) included seventeen leaf morphotypes from the Redoubt Point assemblage in a foliar physiognomy-based paleoclimate reconstruction and concluded that the assemblage grew in the coolest climate of any Paleogene Alaskan flora, with a mean annual temperature of 9°C (48.2°F).

Katmai National Park and Preserve

Katmai National Park and Preserve (KATM) encompasses approximately 4 million acres on the Alaska Peninsula (Figure 1). It is about 160 km (100 mi) northeast of Aniakchak National Monument and Preserve and about 240 km (150 mi) southwest of Lake Clark National Park and Preserve. The park contains a multitude of paleontological resources and has rich Mesozoic marine and Paleogene terrestrial records (Harrington et al., 2023). Fossil plants were first discovered from what is now KATM in exposures at Kukak Bay, now mapped as the Upper Oligocene Hemlock Conglomerate. This collection was made by DeAulton Saunders, a member of the June–July 1899 Harriman Expedition funded by E. H. Harriman, and was studied by the paleobotanist F. H. Knowlton (Knowlton, 1904). Continued exploration of the region has led to the discovery of dozens of other plant fossil localities inside KATM, belonging to the Hemlock Conglomerate and the Paleocene–Lower Eocene Copper Lake Formation (Magoon et al., 1976b; Detterman et al., 1996).

Copper Lake Formation: Fossiliferous exposures of the Copper Lake Formation have been found in the northeastern area of the park near Cape

Douglas and at Dumpling Mountain in the western region of the park (Blodgett et al., 2016). The fossil flora at Cape Douglas is preserved in sandstone and siltstone layers that overly a basal conglomerate. The age of the plant-fossil–bearing horizons is interpreted as Early Eocene on the basis of its taxonomic composition (Detterman et al., 1996, citing a written comm. from J. A. Wolfe). The plant fossils recovered from the Copper Lake Formation at Cape Douglas are leaf compressions and impressions that include three species of ferns, the gymnosperm *Sequoia langsdorfii*, and six species of angiosperms. Taxa named from this flora include the fern *Osmunda dubiosa* and the angiosperms *Myrica banksiaefolia* var. *curta*, *Crataegus alaskensis*, and *Acer douglasense* (Hollick, 1936; Wolfe and Tanai, 1987; Figure 4).

The assemblage of fossil leaves and wood recovered at Dumpling Mountain is also interpreted to date from the Late Paleocene or Early Eocene, and its minimum age is constrained to the Late Eocene based on an intruding diorite dike dated to 39.63 ± 0.17 Ma (Parrish et al., 2010). Twelve distinct leaf morphotypes were identified from 45 collected specimens and include several species – such as *Cercidiphyllum genetrix*, *Glyptostrobus europaeus*, and *Platanus raynoldsii* – that were widespread in Alaska and the western USA during the Paleogene (Parrish et al., 2010). Notably, none of the morphotypes described from Dumpling Mountain represent species identified from the Cape Douglas localities that occur about 160 km (100 mi) to the east, raising the intriguing possibility that the two assemblages represent compositionally distinct but contemporaneous ecological communities. The fossil wood specimens discovered at Dumpling Mountain are well preserved and belong to four species of conifers, including a new species of *Pinus* Section *Parrya*, and the platanaceous wood genus *Platanoxylon* (Parrish et al., 2010).

Hemlock Conglomerate: Fossil plants from the Hemlock Conglomerate have been collected from several localities along the southeastern coast of KATM, including the historical Kukak Bay locality (Knowlton, 1904) and a reference section of the Hemlock Conglomerate measured at Cape Nukshak (Keller and Reiser, 1959). These collections include leaves of ferns, gymnosperms, and angiosperms, as well as conifer and angiosperm reproductive structures (Knowlton, 1904; Hollick, 1936; Wolfe and Tanai, 1987). Common taxa include the gymnosperms *Picea*, *Pinus*, *Sequoia*, and *Taxodium* and the angiosperms *Acer*, *Aesculus*, *Corylus*, and *Alnus* (Figure 4). The collections also include holotypes of 12 taxa (Supplementary Table 7).

Aniakchak National Monument and Preserve

Aniakchak National Monument and Preserve (ANIA), located on the Alaskan Peninsula, was established to preserve Aniakchak Caldera and its surrounding lands (Figure 1). The park contains 600,000 acres and is rich in paleontological resources from the Jurassic, Cretaceous, and Paleogene (Kenworthy and Santucci, 2003). The Paleogene resources include plant fossils of the Tolstoi and Meshik formations (Hollick, 1936; Wolfe, 1980).

Tolstoi Formation: The Tolstoi Formation crops out widely on the Alaska Peninsula, and the age of its strata ranges from Early Paleocene to Middle Eocene (Wolfe, 1980; Detterman et al., 1996). At its type section near Pavlof Bay, about 240 km (150 mi) southwest of ANIA, the Tolstoi Formation represents a warm to subtropical shallow marine environment and contains marine mollusks and dinoflagellates indicative of a Middle Eocene age (Marincovich, 1988; Detterman et al., 1996). At its reference section near Ivanof Bay, about 135 km (85 mi) southwest of ANIA, and in exposures to the northeast (including those in the park), the Tolstoi Formation represents fluvial floodplain and delta sequences and contains a diverse flora indicative of a Paleocene–Eocene age and a warm, wet environment (Wolfe, 1981; Detterman et al., 1996). The plant fossil localities of the Tolstoi Formation within ANIA occur throughout the formation's stratigraphic range and are widely spaced, from the southeastern boundary near Elephant Head Point to the northeastern boundary, where the park borders the Alaskan Peninsula National Wildlife Refuge along Pumice and Painter creeks (Hollick, 1936; Wolfe, 1980; Kenworthy and Santucci, 2003).

In the northeastern area of the park, plant macrofossil localities occur near Painter and Pumice creeks. Sample P-23 was collected by W. R. Smith in 1922 from a mountain southwest of Pumice Creek and yielded the lectotype of the fern *Dryopteris meyeri*, the holotype specimen of *Acer disputabilis* (Figure 4), and other angiosperm leaves assigned to *Ulmus braunii*, *Populus arctica*, and *Zizyphus hyperboreus* (Hollick, 1936). The USGS paleobotanical localities 11577–11580 were discovered in several intervals of a stratigraphic section near Painter Creek and yielded an assemblage including a tree fern, *Cyathea inequalateralis*; a conifer, *M. occidentalis*; and the angiosperms *Pterocarya nigelloides*, *Viburnum variabilis*, and undetermined species of *Acer* and *Platycarya* (Wolfe, 1980). Wolfe (1980) suggested that these localities are Early to Middle Eocene in age and represent the youngest strata of the Tolstoi Formation. Two more localities are present in the central region of the park near Jaw Mountain. Sample P-25 was collected by W. R. Smith in 1922 and includes leaf fossils of the angiosperm family Rhamnaceae (*Z. hyperboreus* and *Paliurus colombi*) and a fossil flower of an extinct species of *Hydrangea* (*Hydrangea alaskana*). Locality USGS 11632 was established near a peak 4.0 km (2.5 mi) south of Jaw Mountain and yielded a leaf belonging to a member of the Menispermaceae (*Cocculus flabella*) and needles of the conifers *Metasequoia* and *Glyptostrobus*. This assemblage is characteristic of the lower Tolstoi deposits and interpreted to be late Paleocene in age (Wolfe, 1980). Exposures of the Tolstoi Formation near Elephant Head Point on Cape Kumlik in the southeastern region of ANIA produced a small collection of plant fossils, which includes undetermined species of *Alnus* and *Vitis* (Wolfe, 1980).

Meshik Formation: The Meshik Formation or Meshik Volcanics is a volcanic unit that contains intervals of volcaniclastic sedimentary rock (Detterman et al., 1981). One such interval, comprising sandstone and water-deposited tuff, is the source of USGS paleobotany locality 11640. These beds, exposed near Aniakchak Crater in the western region of Aniakchak National Monument, have yielded a diverse, angiosperm-rich flora including *Metasequoia, Cercidiphyllum, Quercus, Pterocarya, Ulmus, Cornus*, various Rosaceae, and the holotype specimen of an extinct species of maple, *Acer dettermanii* (USNM 396014) (Wolfe and Tanai, 1987; Wolfe, 1994a). Wolfe (1980) interpreted this assemblage as Middle Eocene in age, an estimate that falls in the early part of the age range suggested by K-Ar dates of the tuff (43.1 ± 1.3 to 24.9 ± 0.49 Ma; Detterman et al., 1996). Wolfe (1994a) used 34 leaf morphotypes from this site to estimate its paleoclimate based on leaf physiognomy and characterized the assemblage as a notophyllous broad-leaved evergreen forest with a mean annual temperature of 13.7°C (56.7°F). Detterman et al. (1981) also reported well-preserved petrified logs, some occurring in original growth position, in exposures of the Meshik Formation on the south shore of Kujulik Bay just south of the southern park boundary.

Kenai Fjords National Park

Kenai Fjords National Park (KEFJ) is located on the Kenai Peninsula in Alaska, preserves about 600,000 acres, and includes the 780 km^2 (300 mi^2) Harding Ice Field, one of four ice caps in North America (Figure 1). Potentially *in situ* logs and woody debris were discovered near Exit Glacier, about 20 km (12 mi) northwest of Seward, Alaska. Initial radiocarbon dating of the woody material from KEFJ indicates an age of 900 years, suggesting that it may represent a forest community buried by an advancing glacier (Kenworthy and Santucci, 2003).

3.1.4 Southeast Alaska Inventory and Monitoring Network

Glacier Bay National Park and Preserve

Glacier Bay National Park and Preserve (GLBA), located in the northeastern part of southeastern Alaska, preserves about 3.3 million acres of land rich in floral and faunal biodiversity and is home to globally recognized tidewater glaciers (Figure 1). Although the area of the park is extensively glaciated, plant macrofossils have been discovered from outcrops of the Oligocene Cenotaph Formation and Miocene–Pleistocene Yakataga Formation on Cenotaph Island, and abundant subfossil Holocene wood has been discovered from Muir Inlet (Figure 3).

Cenotaph Formation: The Cenotaph Formation is an Oligocene volcanic unit interbedded with marine and terrestrial sedimentary rocks that crops out in the Lituya Bay Region, including on Cenotaph Island and on the mainland south of the bay. The plant fossil locality 75-Apr-37 occurs in horizons of the type section of the Cenotaph Formation on Cenotaph Island and has produced angiosperm leaves attributed to *Macclintockia pugetensis*, *Magnolia reticulata*, and undetermined species of *Pterocarya* and the Lauraceae (Wolfe, 1976). *M. pugetensis*, an extinct species of uncertain familial affinity, is restricted to the Kummerian (Oligocene) floral stage in Alaska, and therefore provides strong biostratigraphic correlation with other plant-fossil–bearing units from the Gulf of Alaska (Wolfe, 1977). Plafker (1967) included the Cenotaph Formation within the Topsy Formation, which consists of Middle Oligocene–Upper Miocene shallow marine deposits that are conformably overlain by the Yakataga Formation.

Yakataga Formation: Within GLBA, the Yakataga Formation is exposed along the coast between Lituya Bay and Icy Point, on Cenotaph Island in Lituya Bay, and north and south of Fairweather Glacier (Santucci and Kenworthy, 2008). Exposures on the south shore of Cenotaph Island (USGS paleobotany locality 11186) produced a single leaf fossil assigned to *Fagus* (beech) and are the only source of paleobotanical material from the Yakataga Formation within GLBA. Elsewhere within and beyond the boundaries of GLBA, the Yakataga Formation is known as a marine unit and has produced fossils of foraminifera, bivalves, and gastropods that suggest a Middle Miocene–Pleistocene age (MacNeil et al., 1961; Santucci and Kenworthy, 2008).

Muir Inlet: An early reference to the sub-fossil wood and standing stumps present in what is now called Muir Inlet was made by the naturalist John Muir in the book *Travels in Alaska* (1915). Tree-ring cores and radiocarbon age estimates from these trees have been compiled with records from living trees in GLBA to reconstruct forest succession and paleoclimate in the region over about the past 10,000 years (Lawson et al., 2007; Lawson and Wiles, 2009).

Park Collections: The park collections include three pieces of basal peat deposits and two pieces of petrified wood. The peat was collected from a terrace above Echo Creek north of Lituya Bay and dated to 40,000 ybp. Wood specimens were collected by C. V. Janda from the west side of Adams Valley and from the beach at Dundas Bay and identified as belonging to a conifer and an angiosperm, respectively (Santucci and Kenworthy, 2008).

Figure 5 Map of NPS units with Cenozoic paleobotanical resources in the Pacific West Region, colored by inventory and monitoring network. See Figure 1 legend for details on the park symbology and acronyms.

3.2 Pacific West Region

The Pacific West Region includes 64 NPS units in the states and territories of American Samoa, California, Guam, Hawai'i, Idaho, Nevada, Oregon, and Washington. The parks of this region are clustered in seven inventory and monitoring networks (Figure 5): the Klamath Network (KLMN), the Mediterranean Coast Network (MEDN), the Mojave Desert Network, the North Columbia and Cascades Network, the Pacific Islands Network, the San Francisco Bay Area Network (SFAN), and the Upper Columbia Basin Network. More than 25 geologic formations spanning the Paleocene to Holocene from 20 park units in the Pacific West Region have confirmed paleobotanical resources.

Significant resources from the Pacific West Region include classic paleobotanical localities such as the Eocene Clarno Nut Beds in John Day Fossil Beds National Monument and the Miocene Whitebird site of Nez Perce National Historical Park, as well as a number of lesser-known Plio–Pleistocene sites in the KLMN, MEDN, and SFAN networks of coastal California (Figure 6). The anatomically preserved fruits and seeds of the Clarno Nut Beds represent a phase in the vegetational history of the Pacific Northwest when the flora included a mixture of plants with familial affinities to groups mainly known from tropical

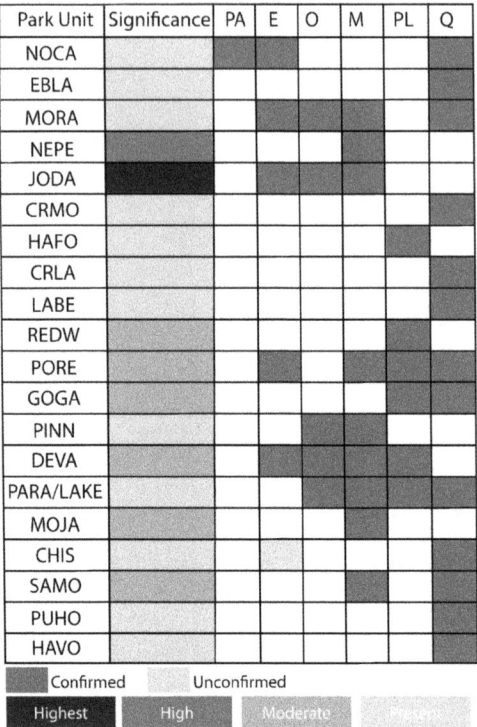

Figure 6 Summary of the temporal distribution and significance of Cenozoic paleobotanical resources in the Pacific West Region. See Figure 2 legend for explanation of significance ratings.

and subtropical areas today (e.g., Icacinaceae, Menispermaceae) as well as to lineages characteristic of temperate North American forests (e.g., Platanaceae). The insight into the Eocene vegetational history of the Pacific West provided by the Clarno Nut Beds is complemented by assemblages preserved outside of NPS units, such as the floras of the Puget Group (Burnham, 1994) and Chuckanut Formation (Mustoe and Gannaway, 1997) in Washington state and the Goshen and Comstock floras of Oregon (see Wing, 1998; Graham, 1999; and DeVore and Pigg, 2010 for reviews of Eocene Pacific Northwest floras).

The Miocene Whitebird site, deposited ca. 30 million years later in western Idaho, includes a greater proportion of broadleaf deciduous angiosperms and conifers characteristic of the extant North American flora, though not necessarily of the Pacific Northwest. Miocene floras from the Columbia Plateau occurring outside of NPS areas are well-documented (Chaney and Axelrod, 1959; Graham, 1965), and recent work has placed them into a stratigraphic context to illustrate changes in floral composition during the Middle Miocene Climatic Optimum (Lowe, 2024).

Fossil plants from the Plio–Pleistocene are preserved in Redwood National Park, Golden Gate National Recreation Area, Point Reyes National Recreation Area, Channel Islands National Park, and Santa Monica Mountains National Recreation Area. These assemblages mainly include genera and species that still occur along the California coast today and have been used to trace the evolutionary history of regional vegetation types such as the California closed-cone pine forest (Axelrod, 1967).

3.2.1 North Coast and Cascades Inventory and Monitoring Network
Northern Cascades National Park Service Complex

The Northern Cascades National Park Service Complex (NOCA) is situated in the state of Washington (Figure 5) and was established to facilitate recreational enjoyment of the North Cascade Mountains and associated glaciers, watercourses, and ecosystems. Cenozoic paleobotanical resources are preserved in NOCA in three contexts: lower Cenozoic sandstones, Quaternary lake sediments, and Quaternary mass wasting deposits.

Chuckanut Formation/Tertiary Sandstones: Sediments of the Chuckanut Formation were deposited in a coastal plain during the Paleocene and Eocene and are well-exposed throughout northwestern Washington and southwestern British Colombia (Figure 7; Mustoe et al., 2007; Fay et al., 2009). Within NOCA, non-diagnostic plant fragments and a fossil log have been found in strata that are variously included within the Chuckanut Formation (Staatz et al., 1972) or within an informal sedimentary unit comprised of sandstones and conglomerates (Tabor et al., 2003). Outside of NOCA, in Chuckanut outcrops near Bellingham, Washington, the formation has produced abundant plant fossils, including palm leaves and foliage of tree ferns, gymnosperms, and woody angiosperms (Mustoe and Gannaway, 1995; Mustoe et al., 2007; Mustoe, 2019). These fossil assemblages are preserved in three different members of the Chuckanut Formation and the stratigraphic succession of plant communities they represent is interpreted to reflect a local transition from a warmer and wetter climate to a dryer and cooler climate, potentially related to the climatic deterioration of the EOT or episodes of local mountain-building (Mustoe and Gannaway, 1997). Well-preserved plant fossils come from fine-grained portions of the Chuckanut Formation, and if such strata are present within NOCA, they are likely stratigraphically close to the fossiliferous sandstone.

Quaternary sediments: Plant macrofossils, pollen, and charcoal have been recovered from Ridley and Thunder lakes and Skagit Valley, all located in the northeastern corner of the park. The cores from Ridley and Thunder lakes

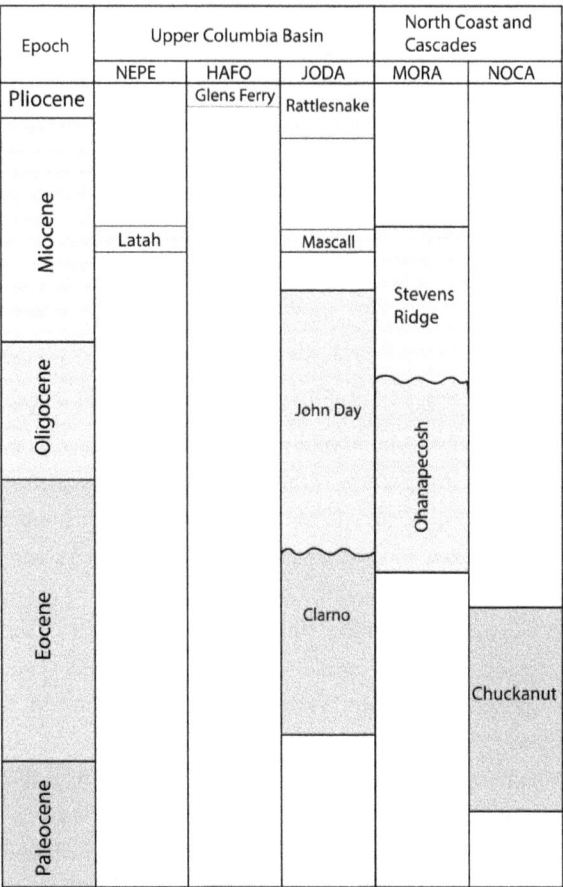

Figure 7 Stratigraphic chart of Tertiary plant-fossil–bearing formations in the Upper Columbia Basin and Northern Coast and Cascades Inventory and Monitoring Networks.

preserve a paleoenvironmental record of the past 14,000 years, and their floras include many plants that inhabit the region today such as *Abies*, *Chamaecyparis*, *Picea*, *Pinus*, *Tsuga*, *Alnus*, *Betula*, and *Fraxinus* (Spooner et al., 2008; Fay et al., 2009). The macrofossils from Skagit Valley document vegetational changes between 24,000 and 17,110 radiocarbon ybp (Riedel, 2007; Fay et al., 2009). Older sediments of this interval are dominated by conifer foliage of *Pinus albicaulis*, *Picea engelmannii*, and *Abies lasiocarpa*, and leaves of *Salix* and fragments of *Selaginella* become more abundant in the younger sediments. Finally, trees buried in growth position are preserved in ca. 7,000-year-old lake deposits that formed after a landslide dammed the Upper Skagit River (Riedel et al., 2001).

Ebey's Landing National Historical Reserve

Ebey's Landing National Historical Reserve (EBLA), located on Whidbey Island of the San Juan Archipelago in Washington state, encompasses about 20,000 acres and includes two state parks and the historic town of Coupeville (Figure 5). The surficial geology of EBLA is dominated by unconsolidated Quaternary sediments that were deposited by glacial and fluvial processes (Fay et al., 2009). The paleobotanical material preserved in these sediments is limited to a few minor occurrences. Sediments accumulated during the Olympia non-glacial interval (ca. 60,000–20,000 ybp) preserve pollen, pinecones, branches, leaf prints, and *in situ* tree roots (Troost, 2002); wood found in an estuary, possibly of Douglas fir, was radiocarbon dated to 1740–1790 or 1810–1960 CE (Polenz et al., 2005a, 2005b).

Mount Rainier National Park

Mount Rainier National Park (MORA) preserves approximately 235,000 acres of land surrounding the peak of Mount Rainier or Tahoma, in the central portion of the Cascade Range of southwestern Washington (Figure 5). The park contains paleobotanical resources in volcaniclastic strata of the Ohanapecosh and Stevens Ridge formations deposited between the Eocene and Miocene and in Holocene lake sediments and mass-wasting deposits.

Ohanapecosh Formation: The Ohanapecosh Formation is comprised of up to 3,000 m (9,800 ft) of volcanic and volcaniclastic sediments within MORA (Fay et al., 2009). The only paleobotanical material yet recovered from these strata within MORA are fragments of wood and leaves, but layers of well-preserved fossil leaves and a forest buried by a lahar deposit are known from correlative strata outside the park. Localities nearby MORA that have yielded significant fossil plant assemblages include a site 600 m (2,000 ft) west of the park boundary on the north side of North Puyallup River, sites on Osborne Mountain and at Catt Creek 3 km (2 mi) southwest of MORA, and a lahar deposit that preserves permineralized wood 24 km (15 mi) west of MORA, near the town of Mineral, Washington (Fay et al., 2009). The leaf floras collected near the park were determined by Roland Brown to include woody angiosperms such as *Alnus*, *Betula*, *Castanea*, *Cercidiphyllum* (katsura tree), *Cinnamomum*, *Laurus*, and *Platanus*, as well as the palm *Sabal*, the gymnosperm *Metasequoia*, and the fern *Thelypteris* (Brown, 1956; Fisher, 1957). The fossil leaves preserved in the lower and upper strata of the Ohanapecosh Formation along Catt Creek are the basis of the Late Eocene–Oligocene age assigned to this unit (Figure 7). Fay et al. (2009) suggested that similar layers may be accessed within MORA near Round Pass and in the saddle just north of Tumtum Peak.

The park collections of MORA contain six paleobotanical specimens that represent conifer needles and wood and angiosperm leaves with a possible affinity to the Fagaceae (Regan Dunn, pers. comm., 2009). Only one of these specimens (MORA 13164) has associated geographic information, and this specimen was recovered in 2006 from the Ohanapecosh River upstream from the Stevens Canyon landslide (Fay et al., 2009). The Ohanapecosh Formation is widely exposed in this drainage basin and is a potential candidate for the source of this specimen.

Stevens Ridge Formation: The Stevens Ridge Formation is predominantly composed of quartz-bearing ash flows and unconformably overlies the Ohanapecosh Formation. A freshwater gastropod (*Viviparus*) of Late Oligocene or Miocene age collected from a quarry about 13 km (8 mi) south of MORA was tentatively correlated with the Stevens Ridge strata exposed in the park (Fiske et al., 1963). Conifer needles are present at several localities of the Stevens Ridge Formation within MORA (Fiske et al., 1963).

Holocene lake sediments and mass wasting deposits: Paleobotanical resources preserved in Holocene lake sediments and mass wasting deposits facilitate the study of recent landscape evolution in MORA by recording local vegetational change and by constraining the age of glacial and tectonic events that lead to mass wasting. Macrofossils sieved from sediment cores taken from three lakes on the south side of Mount Rainier have been used to establish a 6,000-year record of forest succession (Dunwiddie, 1983, 1985, 1986, 1987), and Holocene wood preserved in lahars has been used to constrain the age of numerous postglacial lahars by radiocarbon dating and dendrochronology (Pringle and Scott, 2001; Pringle, 2008).

3.2.2 Upper Columbia Basin Inventory and Monitoring Network

Nez Perce National Historical Park

Nez Perce National Historical Park (NEPE) includes 38 sites in Idaho, Montana, Oregon, and Washington that commemorate the stories and history of the Nimiipuu or Nez Perce people (Figure 5). Six of these sites are on land owned by the National Park Service, and one of these six sites, the Whitebird Battlefield, the location of the first battle of the Nez Perce Flight of 1877, includes exposures of the fossiliferous Latah Formation.

Latah Formation: The classic White Bird locality of the Latah Formation is exposed in a highway roadcut that traverses the Whitebird Battlefield site of NEPE. The Latah Formation is characterized as a sequence of lake beds, ash, sands, and silts interbedded in the Colombia River Basalt, and the formation's fossiliferous strata outcrop extensively in Idaho and eastern Washington

(Kirkham and Johnson, 1929). A tuff dated from the top of the section of the Latah Formation at White Bird provides a minimum age of 14.3 ± 0.1 Ma, but a slightly older age of 15.5–15.8 Ma is likely, based on correlation with the Bully Creek Formation in eastern Oregon (Nash and Perkins, 2012; Figure 7). The paleobotanical resources of the Latah Formation in the vicinity of Whitebird, Idaho, were first documented by Kirkham and Johnson (1929), and their original specimens and localities have been featured in several subsequent studies, notably Berry (1930, 1932, 1934), Ashlee (1932), Brown (1935, 1937), Herendeen and Dilcher (1991), and Manchester et al. (1991). Of the localities mapped by Kirkham and Johnson (1929), stations 8, 9, 12, and 14 (hereafter, White Bird site) are within or adjacent to NEPE, and station 14, which yielded the most species, is probably equivalent to USGS 8444.

Macrofossils from the Latah flora recovered from the White Bird site include attached and isolated foliage, fruits, and seeds that represent 97 species and at least 24 families (Figures 8f and 8g; Kirkham and Johnson, 1929; Ashlee, 1932). The flora is dominated numerically and in number of species by angiosperms; the genera with the greatest representation at White Bird localities are *Acer* (maple), *Quercus*, *Populus*, and *Betula* (Berry, 1932). Gymnosperms from White Bird include *Glyptostrobus*, *Juniperus*, *Pinus* (pine), *Taxodium*, and *Tsuga*, and pteridophytes include *Equisetum*, *Polystichum*, and *Pteris* (Ashlee, 1932). Several genera represented in the paleoflora, including *Cassia*, *Castanea*, *Euonymus*, *Fagus*, *Glyptostrobus*, *Liquidambar*, *Magnolia*, *Malva*, *Nyssa*, *Ostrya*, and *Sophora*, are not found in the region today (Berry, 1932).

The White Bird site has yielded holotype specimens of *Juniperus sabinoides*, *Ceratophyllum praedomersum*, *Aristolochia whitebirdensis*, and *Viburnum whitebirdensis*, and paratypes of *Caesalpinia spokanensis*, *Zizyphoides auriculata*, and *Nordenskioldia interglacialis* (Ashlee, 1932; Herendeen and Dilcher, 1991; Manchester et al., 1991). *C. spokanensis* (Figure 8) includes a fossil fruit collected from White Bird that served as the holotype for the species *Cercis idahoensis*, now a synonym of *C. spokanensis* (Berry, 1930; Brown, 1937; Herendeen and Dilcher, 1991). *Z. auriculata* and *N. interglacialis* are respectively based on leaf and reproductive fossils that co-occur at Miocene localities throughout the Pacific Northwest and are thought to belong to the same taxon (Manchester et al., 1991). Furthermore, these leaves and fruits are, respectively, congeneric with the taxa *Zizyphoides flabellum* and *Nordenskioldia borealis*, which occur in association at Paleocene localities in North America, Greenland, and Norway, and at Paleocene and Eocene localities in eastern Asia (Manchester et al., 1991).

Outside of NEPE, the Latah Formation has produced several other notable fossil assemblages in eastern Washington and western Idaho, including the

Figure 8 Fossil plants from the Pacific West Region. All scales = 1 cm. (a–e) Fossils from the Clarno Nut Beds of John Day National Monument. (a) Holotype specimen of *Anonaspermum bonesii* representing the seed of an extinct member of the tropical plant family Annonaceae. (b) Holotype specimen of *Nyssa scottii*. (c) Holotype of *Laurocarpum nutbedensis* representing the fruit of an extinct member of the tropical plant family Lauraceae. (d) Holotype specimen of *Fortunearites endressii* representing the infructescence of an extinct member of the Hamamelidaceae. (e) Holotype of *Anonmyocarpa ovoideae* representing a fruit of unknown familial affinity. (f) Specimen of *Caesalpinia spokanensis* collected from the Whitebird site of the Latah Formation in Nez Perce National Historic Park. (g) Specimen of *Vitis washingtonensis* from the Whitebird site.

Spokane and Cour d'Alene (Knowlton, 1926), Clarkia (Smiley and Rember, 1985; Rember, 1991), Juliaetta (Kirkham and Johnson, 1929), and Grand Coulee (Berry, 1932) floras. Collectively, these floras document the diversity, composition, and structure of Pacific Northwest forests during the Miocene

Climatic Optimum (Lowe, 2024), and in the case of the Clarkia flora, offer rare insight into the paleobiochemistry of plant organs through the exceptional preservation of leaf cuticular and mesophyll tissues (Logan et al., 1993), as well as pigmentation (Niklas and Brown, 1981), secondary metabolites (Otto et al., 2003), and potentially DNA (Golenberg et al., 1990).

John Day Fossil Beds National Monument

The John Day Fossil Beds National Monument (JODA) in central Oregon consists of three separate protected areas: the Clarno, Painted Hills, and Sheep Rock units. Together they provide access to a succession of Eocene to Miocene fossiliferous strata and volcanic rocks, including the Clarno and John Day formations, the Picture Gorge Basalt, the Mascall Formation, and the Rattlesnake Formation. A geological field trip guide (Dillhoff et al., 2009) provides a useful overview of the local geology and paleontology. The Clarno Unit includes Eocene floras of the Clarno Nut Beds, Hancock (Mammal) Quarry, and Hancock Canyon. The Painted Hills Unit includes several localities of the Early Oligocene Bridge Creek flora studied by Newberry (1898), Lesquereux (1883), Knowlton (1902), Chaney (1927), and Meyer and Manchester (1997). The Sheep Rock Unit, where the National Monument is headquartered, includes the Cant Ranch flora (Meyer and Manchester, 1997) and is adjacent to the Miocene Mascall flora (Knowlton, 1902; Chaney, 1927).

Clarno Formation: The Clarno Formation comprises terrestrial volcanic and volcaniclastic rocks deposited between ca. 54 and 39 Ma (Figure 7) and contains a diverse and well-preserved flora based on wood, leaf, fruit, seed, and palynomorph fossils (Figure 8; Bestland and Retallack, 1994; Manchester, 1994; Wheeler and Manchester, 2002). The flora of the Clarno Formation preserves subtropical and warm temperate fossil plants including horsetails, ferns, cycads, ginkgo, pine, sequoia, palms, banana relatives, and numerous other angiosperms representing both extant and extinct genera.

The Nut Beds site near Hancock Field Station in the Clarno Unit of JODA was so-named by amateur fossil collectors who discovered the location in the 1930s with agatized walnuts and various other nut and seeds preserved as casts, molds, and permineralizations. The fruit and seed flora was initially monographed by Scott (1954), who recognized 10 species, plus six indeterminate additional morphotypes. Continued amateur collecting through the 1970s resulted in greatly expanded collections, such that Manchester (1994) was able to discriminate 145 genera and 173 species representing a warm temperate to subtropical environment. The flora includes numerous extinct genera, but also early records of genera still living today, such as *Juglans, Quercus, Celtis,*

Magnolia, Iodes, Vitis, and *Ampelocissus*. Woody vines are particularly well represented, including multiple genera of the families Menispermaceae, Icacinaceae (e.g., *Iodes*), and Vitaceae (e.g., *Ampelocissus, Vitis*).

The Nut Beds site is unusual in having various plant parts preserved in association with vertebrate remains. Bones and teeth of several mammals and a crocodilian have been recognized (Hanson, 1996). In addition to bones, nuts, and seeds, the site provides the richest known Eocene petrified wood flora worldwide (Wheeler and Manchester, 2002), including 66 genera and 76 species. Fossil leaves are well preserved in the Nut Bed siltstones, with excellent details of venation, but are very fragmentary and time-consuming to collect, requiring gluing of adjacent fragments to obtain complete leaves (Manchester, 1981; Smith, 2023). A survey of dispersed pollen and spores has also been conducted (Smith, 2023).

The Hancock Quarry, also referred to as Mammal Quarry, dates to about 40 Ma, 4 million years younger than the Clarno Nut Beds (Bestland and Retallack, 1994). The quarry was excavated in the 1960s and 1970s for vertebrate fossils and also yielded fruits and seeds (McKee, 1970). The flora includes *Juglans, Vitis,* Icacinaceae, and Menispermaceae. One of the moonseeds was originally treated as *Diploclisia*, a genus today found in Asia, but more recently recognized as *Stephania jacquesii* (Han et al., 2018). *Stephania* is native to the paleotropics and Australasia today (POWO, 2023).

Lahar deposits exposed in Hancock Canyon yield petrified wood, leaves, and occasional fruits. These were investigated by Wheeler and Manchester (2014). Cycad leaves of genus *Ctenis* were found in the Hancock Canyon deposits adjacent to Hancock Field Station as well as in the Nut Beds (Erdei and Manchester, 2015).

John Day Formation: The John Day Formation is composed of tuffs and alluvial deposits and is divided into four members that span from the Late Eocene to the Early Miocene (Kenworthy et al., 2005). The Big Basin Member contains fossiliferous Upper Eocene and Lower Oligocene strata, the latter of which preserve the Bridge Creek flora. The Early Oligocene Bridge Creek flora was first recognized based on collections made by pioneering paleontologist and reverend Thomas Condon from the John Day Formation near the Painted Hills, an area now within John Day National Monument. Two other Bridge Creek localities, Cove Creek and Iron Mountain, are within or immediately adjacent to JODA.

The collections were initially studied by Newberry (1898), Lesquereux (1883), and Knowlton (1902), and subsequently by Chaney (1927) and Meyer and Manchester (1997). The Bridge Creek flora at Painted Hills and Cant Ranch includes a temperate assemblage with *Metasequoia, Alnus, Fagus, Quercus,*

Acer, Tilia, and a few extinct genera such as *Asterocarpinus* (Betulaceae) and *Florissantia* (Malvaceae). In the Cove Creek flora, the taxa represented by the greatest number of specimens are *Alnus* and *Paracarpinus*, a fossil leaf genus associated with *Asterocarpinus*. The Iron Mountain assemblage is similar to the other sites, including abundant fossils of *Metasequoia, Alnus, Acer, Asterocarpinus*, and *Paracarpinus*, but it also includes unique elements such as *Paleophytocrene*, an extinct genus of Icacinaceae; *Zingiberopsis*, a fossil genus of the ginger family (Zingiberaceae); and a relatively high abundance of *Toxicodendron* (Meyer and Manchester, 1997). The Iron Mountain assemblage has the greatest richness of the three sites, preserving a minimum of 44 species, compared to 35 in the Cove Creek assemblage and 38 in the Painted Hill assemblage.

Mascall Formation: The Mascall Formation dates to ca. 15 Ma and comprises rhyolitic tuffs with associated sands and gravels (Bestland and Retallack, 1994). Paleofloras from the Mascall Formation are primarily found outside of JODA, with the exception of unidentified plant fossils reported from the Mascall Overlook in the park's Sheep Rock Unit (Rensberger, 1976; Kenworthy et al., 2005). A number of woody angiosperm and gymnosperm taxa have been recovered from the Mascall Formation near JODA, including *Ginkgo, Sequoia, Quercus, Alnus, Salix, Acer, Ficus*, and *Rhus* (Chaney, 1925; Chaney and Axelrod, 1959).

Rattlesnake Formation: The Rattlesnake Formation is an Upper Miocene unit composed of fanglomerates and is present in the Sheep Rock Unit of JODA. Paleobotanical material is comparatively sparse and includes reports of *Platanus, Ulmus*, and *Salix* fossils (Fremd et al., 1994; Kenworthy et al., 2005).

Craters of the Moon National Monument and Preserve

Craters of the Moon National Monument and Preserve (CRMO) in Idaho was established to preserve a portion of the lava flows in the Craters of the Moon Lava Field (Kenworthy et al., 2005). These lava flows contain abundant tree molds, "lava trees" – upright, cylindrical molds with the impression of bark on the inside – and carbonized wood. More than 140 tree molds have been discovered and mapped in the northern end of the park (Owen and Melander, 2014). The Blue Dragon and Trench Mortar Flat lava flows have the largest number of tree molds, and Stearns (1928) also reported tree molds within CRMO from the "northwest flow from Big Craters," Owl Cavern flow, and Black flow (Stearns, 1928; Owen and Melander, 2014). Carbonized wood was found in some of the tree molds and has been used for carbon-14 dating, revealing the dates of the flows with wood to span from 15,000 to 1,700 ybp

(Bonnichsen and Breckenridge, 1982; Kuntz et al., 1989; Owen and Melander, 2014). Upright lava trees are up to 1.5 m (5 ft) tall and have been found at Trench Mortar Flat, the Tree Mold Flat lava flows, and the wilderness area of CRMO (Stearns, 1928).

Hagerman Fossil Beds National Monument

The Hagerman Fossil Beds National Monument (HAFO) in Hagerman, Idaho, is renowned for its late Pliocene vertebrate faunas that include the "Hagerman Horse" (*Equus simplicidens*), the earliest recognized member of the extant genus *Equus* (Kenworthy et al., 2005). Paleobotanical material collected from HAFO is limited to pollen and rare occurrences of fossil wood.

Glenns Ferry Formation: The Pliocene Glenns Ferry Formation represents deposition in lacustrine, floodplain, and fluvial environments and contains basalt flows, silicic volcanic ash, and basaltic pyroclastic deposits that constrain the age of the unit to the Pliocene (ca. 5–3 Ma) (Kenworthy et al., 2005). Fossil wood has been occasionally found in the Glenns Ferry Formation but has not been formally studied. One well-preserved specimen showing anatomical detail was reported in the National Park Service Morning Report for December 23, 1993 (Kenworthy et al., 2005). Palynomorphs isolated from sediments of this unit reveal a community of conifers with sparse hardwoods and many steppe elements in the Chenopodiaceae and Asteraceae (Leopold and Denton, 1987). The palynomorph assemblage becomes increasingly rich in grass pollen over time and, alongside the coeval vertebrate fauna, documents the transition of this region from forests to an open steppe and grassland environment (Leopold and Denton, 1987).

3.2.3 Klamath Inventory and Monitoring Network

Crater Lake National Park

Crater Lake National Park (CRLA) in the Cascade Range of southwestern Oregon preserves a caldera lake that formed after the eruption and collapse of Mount Mazama 7,700 ybp. Carbonized tree stumps are preserved in Pleistocene glacial till deposits, and carbonized wood is present in Holocene pumice fall deposits and ignimbrite deposits (Santucci and Kenworthy, 2009). The estimated age of the eruption that produced Crater Lake is based on radiocarbon dating of charcoal twigs preserved in ignimbrite (Bacon, 2008).

Lava Beds National Monument

Lava Beds National Monument in the Cascade Range of northern California preserves flows that contain tree molds, charcoal, and charred wood. Three such flows have been identified inside the monument, namely the Valentine Cave (2,260 ybp), Black Crater (3,080 ybp), and Callahan flows (1,120 ybp) (KellerLynn, 2014). Charred wood and charcoal from these flows were used for radiocarbon dating (KellerLynn, 2014).

Redwood National Park and State Parks

Redwood National Park (REDW) was established to preserve temperate rainforests along the Pacific coast of northern California that include the world's tallest trees, the coast redwood (*Sequoia sempervirens*). The park encompasses four units – Redwood National Park, and Jedediah Smith, Del Norte, and Prairie Creek state parks – that are cooperatively managed by the state of California and the federal government. The Prairie Creek State Park unit of REDW contains a Late Pliocene flora based on tree and root casts, logs, leaf impressions, carbonized leaves, cones, and peat (Criswell, 1968; Trexler, 1989).

Prairie Creek Formation: Plant fossils of the Prairie Creek Formation have been recovered from at least six horizons within a 244 m (800 ft) section of conglomerate, sandstone, and siltstone exposed on the coastal cliffs of the Gold Bluff area within Prairie Creek State Park (Santucci and Kenworthy, 2009). The sequence is interpreted to have been deposited during the Late Pliocene and Early Pleistocene (Figure 9) in a fluvial floodplain of the ancestral Klamath River, and its fossil assemblage comprises eighteen species of plants, an unidentified fossil vertebrate, and two species of freshwater pelecypods (Criswell, 1968). The Prairie Creek flora includes the liverwort *Marchantia*, the fern *Dryopteris*, herbaceous and woody angiosperms – including *Acer*, *Alnus*, and *Betula* – and gymnosperms, including *S. sempervirens*. Casts of trees infilled with sand and associated with carbonized wood are also present but have not been identified (Trexler, 1989).

3.2.4 San Francisco Bay Area Inventory and Monitoring Network

Point Reyes National Seashore

Point Reyes National Seashore (PORE) is situated along the coast north of San Francisco, California, and was established to preserve about 70,000 acres of beaches, sea cliffs, coastal dunes, grasslands, and Douglas fir forests (Elder et al., 2008). Plant fossils have been noted from several formations exposed within PORE, including the Merced, Millerton, and Olema Creek formations that have also yielded plant fossils in the Golden

Epoch	San Francisco Bay Area			Mediterranean		Klamath
	GOGA	PORE	PINN	CHIS	SAMO	REDW
Pleistocene		Olema Creek		Unnamed sediments	Unnamed sediments	
		Millerton				
Pliocene		Merced				Prairie Creek
Miocene		Purisma				
					Modelo	
					Conejo Volcanics	
					Topanga Canyon	
			MIOLpd		Vaqueros	

Figure 9 Neogene plant-fossil–bearing formations in the Klamath, Mediterranean Coast, and San Francisco Bay Area Inventory and Monitoring Networks.

Gate National Recreation Area (see below), but few of these occurrences have been described in detail.

Point Reyes Conglomerate: The Eocene Point Reyes Conglomerate is exposed on the western portion of the Point Reyes Peninsula and is the oldest geologic unit exposed within the boundaries of the national seashore. The conglomerate contains carbonized plant remains and seeds in silty micaceous sand horizons (Galloway, 1977).

Purisma Formation: The Upper Miocene–Pliocene Purisma Formation (ca. 6.7–4 Ma) is the uppermost unit in a sequence of three formations, which as a whole grade from glauconite-bearing greensands upward to shales and white siltstone and mudstones (Galloway, 1977; Elder et al., 2008). Axelrod (1971) described pinecones collected from strata of the Purisma Formation exposed along Drakes Bay within PORE.

Olema Creek Formation: The Pleistocene Olema Creek Formation is composed of interbedded estuarine and alluvial sediments deposited on a coastal plain, and its section commonly includes fine-grained overbank and backswamp deposits with abundant carbonaceous material (Grove et al., 1995; Figure 9). The backswamp deposits are interpreted as such based on the occurrence of peat beds that suggest the presence of standing water. The peat beds preserve plant macrofossils and pollen. The macrofossil assemblage includes unidentified leaves, roots, branches, and cones, and palynomorphs include aquatic and riparian taxa such as *Potamogeton*, Cyperaceae, *Salix*, and *Alnus* (Grove et al., 1995). Thermoluminescence dating constrains the age of the Olema Creek Formation deposits of Tomales Bay to about 130,000 ybp (Grove et al., 1995).

Golden Gate National Recreation Area

The Golden Gate National Recreation Area (GOGA) encompasses about 80,000 acres in the vicinity of San Francisco, California, with exceptional exposures of late Cenozoic marine and terrestrial rocks that include three paleobotanically productive formations (Henkel et al., 2015; Figure 9).

Merced Formation: The Merced Formation includes terrestrial and marine deposits from the late Pliocene to Pleistocene. Fossil pinecones from sandstones of the Merced Formation exposed in eroding sea cliffs at Mussel Rock and Thornton Beach were first collected and studied in the early 1930s (Dorf, 1933; Mason, 1934), and pinecones from Thornton Beach have been collected as recently as 2012 (Henkel et al., 2015). Collections from a site (locality 159 in Dorf, 1933) 3 km (2 miles) north of Mussel Rock produced paratype specimens of *Pseudotsuga sonomensis* and *Pinus masoni* that appear to be closely related to extant *Pseudotsuga taxifolia* and *Pinus muricata* (Dorf, 1933). Strata unconformably underlying the Merced Formation from another site near Mussel Rock produced the type specimen of *Pinus lawsoniana*, implying a minimum Pliocene age for this species (Axelrod, 1967). Routine monitoring of these fossiliferous and actively eroding sea cliffs should be a priority for paleontological resource management in GOGA.

Millerton Formation: The Millerton Formation comprises Upper Pleistocene marine and freshwater clays, silts, sands, gravels, and conglomerate (Henkel et al., 2015). Within GOGA, the formation is exposed on the shore of Tomales Bay and has yielded a rich floral record from outcrops on the eastern side of the bay (Elder et al., 2008). The floral assemblage collected from the Millerton Formation includes unaltered or partially carbonized wood, seeds, fruits, and flowers that represent more than fifty species (Weaver, 1949; Johnson, 1962), and at least two dozen species characteristic of modern closed-canopy pine forests (Axelrod, 1971). Richards and Thurber (1966) constrained the age of the Millerton Formation on the east coast of Tomales Bay to older than 50,000 ybp based on carbon-14 and thorium-230–uranium-234 dating of mollusk shells.

Olema Creek Formation: The Pleistocene Olema Creek Formation is stratigraphically above the Millerton Formation and preserves peat, woody debris, upright tree trunks, and aquatic and riparian plant material within GOGA (Galloway, 1977; Grove et al., 1995; Henkel et al., 2015). See the Point Reyes National Seashore section for further discussion on the paleobotany of this formation.

Pinnacles National Monument

Pinnacles National Monument, located between the San Benito and Salinas valleys in California, was established to protect its unique landscape formed by rock spires, crags, and pinnacles of volcanic breccia. Because the geology of the unit is predominantly composed of metamorphic and plutonic rocks, past and future collection of paleobotanical material is limited. The only paleobotanical material reported from the park to date are clasts in dacite breccia that resemble fossil wood. These clasts were found in an unnamed Oligocene–Miocene volcanic unit (MIOLpd of Barthelmes, 2022).

3.2.5 Mojave Desert Inventory and Monitoring Network

Death Valley National Park

Death Valley National Park (DEVA) occupies more than 3 million acres in eastern California and western Nevada and is the largest unit managed by the NPS in the contiguous USA. The park preserves a remarkably complete geologic record of two time periods: the Neoproterozoic to the middle Permian (ca. 790–270 Ma) and the Late Eocene (ca. 40 Ma) to present (Tweet et al., 2016). The Cenozoic strata of DEVA are intermixed with igneous flows and volcanic ash falls and were mainly deposited in small, extensional basins that form the characteristic basin and range topography of the region today. Minor Cenozoic paleobotanical resources occur in the Upper Eocene–Lower Oligocene Titus Canyon Formation and the Upper Miocene to Pliocene Furnace Creek and Copper Canyon formations.

Titus Canyon Formation: The Upper Eocene–Lower Oligocene Titus Canyon Formation is exposed in the Grapevine and Funeral mountains in the eastern region of the park. Petrified wood and a partial pinecone identified as *Pinus crossii* have been recovered from Titus Canyon sediments within DEVA (Axelrod, 1986). The age of these sediments is constrained to 34.7–30.4 Ma by radiometric dating of tuffs from the base of the unit and the overlying formation (Snow and Lux, 1999; Lander et al., 2022).

Furnace Creek and Copper Canyon formations: The Furnace Creek and Copper Canyon formations comprise volcanic basalts and fluvial and lacustrine sediments deposited from the Late Miocene into the Pliocene (ca. 5.87–3.15 Ma). These strata preserve leaves of palms, *Lyonothamnus* (a genus in the Rosaceae now restricted to the Channel Islands), indeterminate grass-like plants, reed casts, charophytes, and a coprolite containing gymnosperm and angiosperm pollen (Spurr, 1903; Axelrod, 1940; Nyborg and Santucci, 1999; Tweet et al., 2016).

Grand Canyon–Parashant National Monument and Lake Mead National Recreation Area

Grand Canyon–Parashant National Monument (PARA) and Lake Mead National Recreation Area (LAKE) are geographically adjacent and share the same stratigraphy. The oldest National Recreation Area in the country, LAKE encompasses approximately 1.5 million acres in southeastern Nevada and northwestern Arizona. To protect a remote area of northwestern Arizona with exceptional Paleozoic paleontological resources, PARA was established in part from land formerly included within LAKE (Tweet et al., 2021). Minor occurrences of paleobotanical resources are known from each NPS unit and are derived from several different geologic formations.

Rainbow Gardens Formation: The Oligocene–Miocene Rainbow Gardens Formation (ca. 25–17 Ma) crops out extensively in the northern portion of LAKE and at Tassi Springs in PARA (Lamb et al., 2018; Tweet et al., 2021). This formation is composed of fluvial siliciclastic and paludal and lacustrine carbonate rocks and preserves plant stems, root traces, and unclassified "tubular" plant casts within the parks (Longwell, 1936; Beard, 1996; Howard et al., 2003).

Other Reports: Lake Mead hosts some paleobotanical resources that are not present in PARA. These include a silicified palm stem recovered from unidentified Paleogene sediments, a floral assemblage from the Upper Miocene (ca. 12–6.2 Ma) Hualapai Limestone, and sub-fossil and fossil wood from the Plio–Pleistocene Bullhead Alluvium (Howard et al., 2015; Bonde et al., 2018). The latter unit was formerly considered a member of the Muddy Creek Formation but now constitutes its own formation (Bohannon, 1984). Near the Temple Bar area of LAKE, the Hualapai Limestone has produced impressions of stems, roots, and leaves, some of which resemble grass or reed-like vegetation, and a possible *Equisetum* stem (Blair and Armstrong, 1979).

Mojave National Preserve

Mojave National Preserve (MOJA), located in southern California about 50 km (30 mi) southeast of DEVA, was established to preserve approximately 1.6 million acres of the Mojave Desert. The Cenozoic strata exposed in

MOJA are of Neogene age and comprise alluvial fan, eolian, fluvial, and lacustrine sediments interfingered with volcanic flows and eruptions (Tweet et al., 2016; Johnson et al., 2023). Paleobotanical material has been recovered from a few sites within the park, most notably from lacustrine sediments of Wild Horse Mesa and Hackberry Wash that are informally named the Winkler formation (Reynolds et al., 1995). Sediments of the Winkler Formation are between two volcanic tuffs dated at 18.5 Ma and 17.7 Ma and preserve logs and wood of *S. langsdorfii* as well as other unidentified conifer wood and needles (Hazzard, 1954; Reynolds et al., 1995; Santucci et al., 2004). These floral remains are preserved alongside coeval ostracods, flamingo-like tracks, and a mammalian fauna that includes the rhinoceros *Menoceras* (Reynolds et al., 1995).

Other paleobotanical records from MOJA include reports of petrified "reeds" or reed casts from Hackberry Wash (Strong 1975), palm wood and other unclassified wood from Hackberry Mountain (Strong, 1975), unclassified petrified wood from the Providence Mountains, and *Sequoia* wood recorded from Barber Canyon and the mouth of Beecher Canyon (Reynolds et al., 1995; Tweet et al., 2016). Improved documentation of the stratigraphic context of these occurrences of petrified wood will clarify whether they are derived from one or more geologic units and may aid in the protection of this resource.

3.2.6 Mediterranean Coast Inventory and Monitoring Network
Channel Islands National Park

Channel Islands National Park (CHIS) preserves the unique ecological, archeological, and geological resources of five of the eight Channel Islands that comprise the Channel Island Archipelago off the coast of southern California. The islands included in CHIS are the four northern islands – San Miguel, Santa Rosa, Santa Cruz, and Anacapa – as well as the northernmost island of the Southern Channel Islands, Santa Barbara. Major paleontological discoveries from CHIS are invertebrate faunas and remains of the pygmy mammoth (*Mammuthus exilis*), but the islands also preserve a wide range of paleobotanical resources. The resources include a possible *Cercidiphyllum* seed from the Eocene Cozy Dell Formation, root casts in the Pleistocene Potato Harbor Formation, a Pleistocene macroflora on Santa Cruz Island, seeds and caliche root and tree casts from Pleistocene deposits on San Miguel Island, and wood of *Cupressus goveniana*, *Pinus* and *Prunus* from Quaternary sediments on Santa Rosa Island (Tweet et al., 2012b, 2020).

San Miguel Island Caliche Deposits: Pleistocene caliche (pedogenic calcium carbonate) deposits are present on several of the Channel Islands and are particularly well developed on San Miguel Island, where carbonate cement formed molds and casts on plant roots and trunks. These caliche deposits appear as pseudo-fossil forests and are thought to have formed during episodes in which windblown sand buried the standing vegetation. The original components of the standing vegetation are uncertain, but some workers have suggested that caliche concretions formed around roots of *Rhus integrifolia* and that the buried vegetation was shrubland (Cockerell, 1938; Johnson, 1977). In addition to the calcified stumps and roots, plant macrofossils are also preserved as charcoalified stumps; cone and leaf impressions in calcareous tufa in late Quaternary sediments on San Miguel Island include *Pinus* and *Ribes* (Johnson, 1977).

Santa Cruz Island: A rich assemblage of seeds, cones, needles, leaf fragments, stem fragments, wood, and logs has been recovered from a clay, sand, and gravel deposit in the valley of Willow Creek in Cañada de los Sauces del Oeste on the southwestern part of the island (Chaney and Mason, 1930; Tweet et al., 2020). Radiocarbon dates of Douglas fir and charcoal found in this deposit constrain the age to 39,010–11,760 ybp (Fergusson and Libby, 1963). The assemblage includes mosses, gymnosperms, and woody and herbaceous angiosperms. Mosses are represented by three species; gymnosperms are represented by wood, foliage, and cones of *C. goveniana*, *Pseudotsuga menziesii*, and two species of *Pinus;* angiosperms include seeds of the genera *Arctostaphylos*, *Arceuthobium*, *Ceanothus*, *Cornus*, *Garrya*, and *Myrica* (Chaney and Mason, 1930; Johnson, 1977).

Santa Monica Mountains National Recreation Area

Santa Monica Mountains National Recreation Area (SAMO) protects ca. 150,000 acres of the Santa Monica Mountain range on the coast of southern California. The recreation area borders Los Angeles to the southeast and is the largest urban national park in the USA. The coast, canyons, and mountains of the park contain a Mediterranean-climate ecosystem (chapparal) and extensive archeological and paleontological remains. The Neogene fossiliferous rocks of SAMO (Figure 6) were deposited in coastal floodplains and nearshore marine and submarine environments, and the abundance and quality of paleobotanical material preserved in these rocks varies considerably.

Unidentified fragments, wood, and root traces: Several formations within SAMO contain paleobotanical resources that are too fragmentary, poorly preserved, or biologically uninformative to be placed in a detailed ecological or evolutionary context. The Paleocene Simi Conglomerate comprises sandstones and conglomerate deposited in fluvial, nearshore marine, and submarine environments and is known to contain plant fragments at Runyon and Solstice canyons (Colburn et al., 1988) as well as fossil wood (Parker, 1983). The Sespe Formation is a predominantly terrestrial unit that was affected by a marine transgression toward the end of its deposition (Tweet et al., 2012b). The formation includes root casts in its Piuma Member (Lander, 2011) and mammals that range in age from Uintan (46.2–42.0 Ma) to Hemingfordian (20.6–16.3 Ma). The Middle Miocene Calabasas Formation, also known as the Upper Topanga Formation, is an offshore marine unit that contains locally abundant plant casts in the northern Santa Monica Mountains (Yerkes and Campbell, 1979).

Vaqueros Formation: The Vaqueros Formation represents sediments deposited in deltaic and marine environments during a period of sea level rise in the Late Oligocene. The unit unconformably overlies the terrestrial Sespe Formation in its western exposures and interfingers with fluvial deposits of the Sespe Formation in its eastern exposures. Fossil palm fronds (*Sabalites*), fossil log compressions, mollusks, and cartilaginous fish tooth fragments are preserved in a road cut along Piuma Road within SAMO that exposes gray marine sandstone of the Carbon Canyon Tongue of the Vaqueros Formation, which occurs between the lower and upper Piuma Members of the Sespe Formation. The age of these materials is constrained by $^{40}Ar/^{39}Ar$ dates of tuff in the upper strata of the lower Piuma Member (27.2 ± 0.2 Ma) and the lower strata of the upper Piuma Member (21.1 ± 0.2 Ma) (Lander 2011).

Topanga Canyon Formation: The Topanga Canyon Formation is divided into four members that span the Early to Middle Miocene (ca. 23.0–15.5 Ma), and collectively their sediments are interpreted to represent two episodes of transgression separated by a prograding delta (Tweet et al., 2012b). Within SAMO, paleobotanical material has been reported from the formation's basal, deep-marine Encinal Member, and from the Fernwood Member, a terrestrial unit that lies between the nearshore Saddle Peak and Cold Creek members. The Encinal Member contains plant and wood fragments (Flack, 1993), and the Fernwood Member contains leaves of *Aralia* and petrified wood of *Platanoxylon* (Lander, 2011).

Conejo Volcanics: The Conejo Volcanics are comprised of a sequence of pillow lavas, pillow breccias, laharic breccias, and breccia conglomerates that represent a transition between submarine and subaerial volcanic deposition (Stadum and Wiegand, 1999). Sparse vascular plant fossils have been recovered from lower strata of this unit exposed within SAMO (Stanton and Alderson, 2013), and a notable assemblage of fossil wood was discovered in a tuff breccia in the unit's upper strata exposed near Westlake Village, adjacent to SAMO (Stadum and Wiegand, 1999). The assemblage includes several well-preserved specimens, and analysis of thin sections revealed the presence of the conifers *P. menziesii*, *Calocedrus decurrens*, and *Pinus ponderosa*, as well as angiosperm wood of the Fagaceae, Rhamnaceae, and Rosaceae (Stadum and Wiegand, 1999).

Modelo Formation: The Modelo Formation is composed of shale, sandstone, conglomeratic sandstone, and volcanic ash deposited in a marine environment during the Middle to Late Miocene (Tweet et al., 2012b). The formation preserves abundant foraminifera that constrain its age to ca. 15–5.2 Ma (Clark, 1983; Rumelhart and Ingersoll, 1997). Paleobotanical resources of the Modelo Formation within SAMO includes leaves of *Platanaus dissecta* recovered from a foraminiferal platy shale layer in the unit's lower strata (Hoots, 1931), small pieces of petrified wood (Soper, 1938), and algae from Sepulveda Canyon (Parker and Dawson, 1965). The algae collection includes holotype specimens of brown and red algae (Tweet et al., 2012b; Supplementary Table 7).

Unnamed Pleistocene sediments: Pleistocene pinecones of *Pinus radiata* and *Pinus remorata* and an *Arctostaphylos* seed have been recovered from the mouth of Potrero Canyon in SAMO (Axelrod, 1967, 1990). Axelrod (1967) included the cones found in Potrero Canyon Park in a study that traced the evolutionary history of modern Californian closed-cone pines back to the Miocene. Today, *P. radiata* persists only as five disjunct populations on the California coast and has become critically endangered due to habitat fragmentation, land conversion, and native pathogens (Millar, 1998).

3.2.7 Pacific Island Inventory and Monitoring Network

Pu'uhonua o Hōnaunau National Historical Park

Pu'uhonua o Hōnaunau National Historical Park (PUHO) was established to preserve 182 acres on the shoreline of the Mauna Loa volcano, on the western coast of the island of Hawai'i, which contain several significant archeological sites. Lava flows in PUHO have produced tree molds similar to those preserved in Hawai'i Volcanoes National Park. Nearly three dozen molds have been identified and are dated to about 1,000 ybp (Woodcock and Kalodimos, 2003, 2005). Although detail on many of these molds was lost due to "squeeze-up" formations, in which lava breached the mold from below and infilled portions of the hollow molds, they appear to represent *Prichardia*, a native palm that still grows in the park today (Hunt et al., 2007).

Hawai'i Volcanoes National Park

Hawai'i Volcanoes National Park, on the island of Hawai'i, includes the peaks of Kīlauea and Mauna Loa, two of the world's most active volcanoes. In 1962, an eruption of lava through the floor of Aloi Crater temporarily filled the crater with a lava lake that reached depths of up to 50 feet and engulfed a stand of *Metrosideros collina* var. *polymorpha* growing on the crater floor. A subsequent eruption lower down on the mountain caused the lava lake to recede into a vent, leaving tree molds and casts in lava pillars, alongside the remains of trees that had not completely burned. The casts often preserve features such as leaf attachment and bark imprints (Hamilton, 1964; Hunt et al., 2007).

3.3 Intermountain Region

The Intermountain Region includes 62 NPS units in Arizona, Colorado, Montana, New Mexico, Oklahoma, Texas, Utah, and Wyoming. This region is further divided into seven inventory and monitoring networks (Figure 10): the Chihuahuan Desert Network, Greater Yellowstone Network, the Northern Colorado Plateau Network, the Rocky Mountain Network, the Southern Colorado Plateau Network, the Sonoran Desert Network, and the Southern Plains Network. Paleobotanical material has been discovered in 15 NPS units that fall within this region and along two National Heritage Trails that pass through the region.

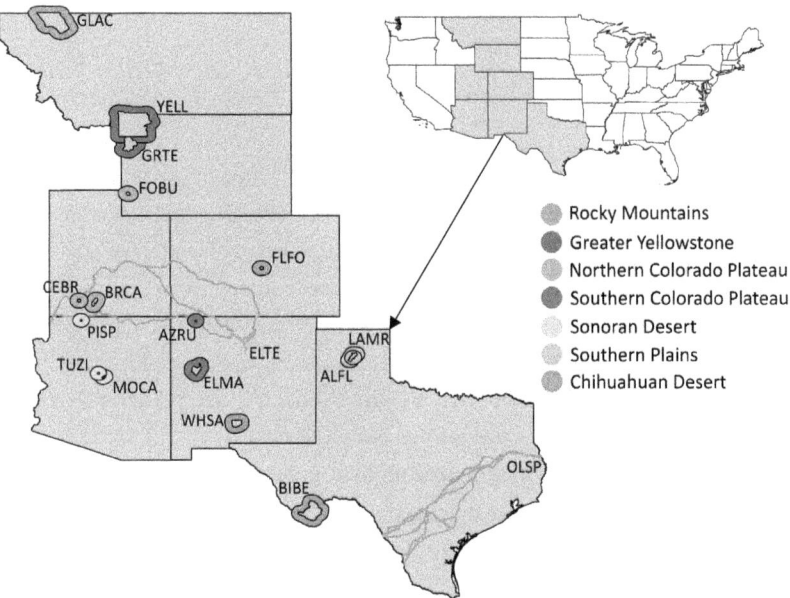

Figure 10 Map of NPS units with Cenozoic paleobotanical resources in the Intermountain Region, colored by inventory and monitoring network. See Figure 1 legend for details on the park symbology and acronyms.

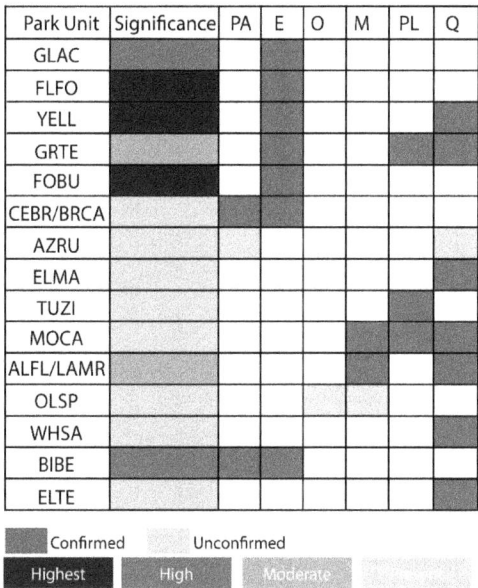

Figure 11 Stratigraphic chart of Paleogene plant-fossil–bearing formations in the Intermountain Region.

The Intermountain Region is particularly rich in Eocene paleobotanical resources, which are conserved in eight NPS units (Figures 11 and 12). Among these resources are the abundantly fossiliferous shales of Florissant Fossil Beds National Monument and Fossil Butte National Monument that preserve some of the best-known Eocene plant assemblages worldwide, and the iconic fossil forests of Yellowstone National Park that document an interesting mode of fossil preservation. The long history of collection and study of fossil plants from Florissant and Fossil Butte has enabled detailed reconstruction of their fossil species (e.g., Manchester and Crane, 1987; McMurran and Manchester, 2010), ecological communities (e.g., Meyer, 2003; Grande, 2013), plant–insect interactions (e.g., Smith, 2008), and surrounding landscape (e.g., Meyer, 1986). Importantly, these floras also serve as reference points for the identification of fossil taxa and interpretation of fossil plant communities preserved in other Paleogene sediments of the western USA and feature prominently in some syntheses of the vegetational history of North America (e.g., Wing, 1987, 1998; Graham 1999; DeVore and Pigg, 2010; Pigg and DeVore, 2010). The fossil plants of Yellowstone also have a long history of collection and similar potential to serve as a key reference point in the Eocene history of the Rocky Mountain Region but are not nearly as well-studied as the assemblages from Florissant or Fossil Butte

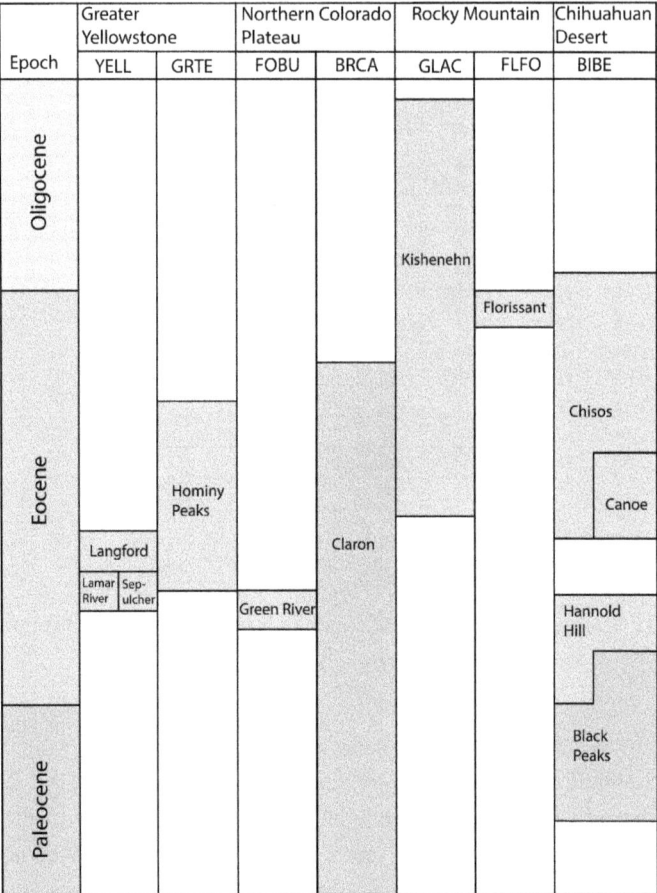

Figure 12 Summary of the temporal distribution and significance of Cenozoic paleobotanical resources in the Intermountain Region. See Figure 2 legend for explanation of significance ratings.

(Lesquereux, 1872; Knowlton, 1899; Dorf, 1964). The flora of the Lamar River and Sepulcher Formations dates from the Early Eocene Climatic Optimum and is coeval with several other Eocene floras of Wyoming (Wing, 1987) but has not been revised since its initial description (Knowlton, 1899; Supplementary Table 8).

3.3.1 Rocky Mountain Inventory and Monitoring Network

Glacier National Park

Glacier National Park (GLAC) encompasses over a million acres in far northern Montana and shares its northern boundary with the US–Canada border. The

Kishenehn basin, a structural and depositional basin spanning approximately 150 km (93 mi) along the Flathead Fault between northern Montana and southern British Columbia, is a prominent geologic feature within GLAC and preserves diverse floral and faunal remains in the Middle Eocene Kishenehn Formation.

Kishenehn Formation: Fossiliferous exposures of the Middle Eocene Kishenehn Formation are known from two regions of the Kishenehn basin: the North Fork of the Flathead River Region in southern British Columbia, Canada, and the Middle Fork of the Flathead River in northern Montana. In the Middle Fork Region, which includes the southern boundary of GLAC, the Kishenehn Formation is divided into the lower Coal Creek Member and the upper Pinchot Conglomerate. The Coal Creek Member is estimated to be 46.2 ± 0.4 Ma based on $^{40}Ar/^{39}Ar$ analysis and preserves a diverse flora of more than 108 plant taxa (Smith et al., 2023) alongside a fauna that includes insects (16 orders and dozens of families; Greenwalt et al., 2014), mammals (26 species; Dawson and Constenius, 2018), mollusks (76 species; Pierce and Constenius, 2001, 2014), and fish (Constenius et al., 1989).

Plant fossils of the Kishenehn Formation are found as compressions and impressions of leaves, cones, fruits, and seeds in lacustrine oil shales and siltstones. According to Smith et al. (2023), the cones, fruits, and seeds include representatives of at least 14 families: Cupressaceae, Pinaceae, Betulaceae, Brassicaceae, Cercidiphyllaceae, Eucommiaceae, Juglandaceae, Oleaceae, Platanaceae, Rutaceae, Salicaceae, Sapindaceae, Simaroubaceae, and Ulmaceae (Figure 13). Vegetative material includes stems of *Equisetum* and at least 27 types of leaves. The most common of these leaf types are *Alnus*, *Macginitiea wyomingensis*, *Trochodendroides*, and *Metasequoia*. The flora also includes several aquatic plants: *Lemna*, *Wolffiella*, *Potamogeton*, and *Ceratophyllum*.

The Kishenehn flora is significant for its geographic placement relative to other Eocene floras, its rich diversity of vegetative and reproductive material, and its cooccurrence with a diverse faunal assemblage. The sites within GLAC are the northernmost Middle Eocene paleobotanical localities in the US Rocky Mountain Region, and highly diverse. The flora contains many taxa that today grow naturally only in Asia (e.g., *Ailanthus*, *Koelreuteria*, *Boniodendron*, *Dipteronia*, *Engelhardia* [*Palaeocarya*], and Cercidiphyllaceae). The flora also contains fossils that represent the earliest known occurrences of the genera *Lemna*, *Thalictrum*, *Myriophyllum*, *Thlaspi*, and *Chenopodium*, and the first reported fossil occurrences of *Wolffiella* and *Agrimonia*. Many taxa have unclear affinities to any particular plant family, including two new genera and species established by Smith et al. (2023).

Figure 13 Fossil plants from the Intermountain Region. All scale bars = 1 cm. (a–c) Fossil leaves and reproductive structures collected from Big Bend National Park. (d) Leaf of *Diospyros lamarensis* collected from the banks of the Lamar River in Yellowstone National Park. (e) Holotype specimen of *Devallia montana* collected from Amethyst Mountain in Yellowstone National Park. (f) *Paleobowdichia lamarensis* collected from Yellowstone National Park. (g) Unidentified angiosperm leaves collected from the Claron Formation in Bryce Canyon National Park. (h) *Fagopsis longifolia* from Florissant Fossil Beds. (i) *Plagiopodopsis cockerelliae* from Florissant Fossil Beds. (j) Winged seed of *Carpolithes lunatus*, an extinct member of the Cercidiphyllaceae, recovered from the Kishenehn Formation in Glacier National Park. (k) Fruit of *Thlaspi*, an extant genus in the Brassicaceae, recovered from Glacier National Park.

Furthermore, there are paleobotanically productive horizons of the Kishenehn Formation within GLAC that are yet to be explored. In addition to the highly fossiliferous shales, Kishenehn sandstones are known to contain at least one taxon, *Macginitiea*.

Florissant Fossil Beds National Monument

The paleobotanical resources now protected at Florissant Fossil National Monument (FLFO) were first discovered by European scientists in the early 1870s, long before it was authorized as a National Monument in 1969. The monument was created in opposition to housing development plans (Leopold and Meyer, 2012), with the intent to continuously monitor the fossil resources, develop interpretive programs, and aid scientists in the study and explication of its natural history. Consequently, the fossil resources of FLFO are among the best known and most secure of any NPS unit, as can be evidenced by the Florissant fossil database (https://florissantdatabase.colorado.edu/) and the mountain of scientific journal articles and popular scientific literature pertaining to the Florissant Lake Beds that have been published since its incorporation (Meyer et al., 2008).

Florissant Formation: All plant macrofossils recovered from FLFO come from the Florissant Formation. The Florissant Formation contacts the Wall Mountain Tuff at its base and lava flows of the Thirtynine Mile Volcanic field at the top (Evanoff et al., 2001). This formation comprises shales, mudstones, and conglomerate that were deposited in the Florissant paleovalley. The valley was episodically dammed by volcaniclastic debris, giving rise to the formation's highly fossiliferous lacustrine deposits known as the Florissant Lake Beds (Evanoff et al., 2001). Outcrops of these ancient lakebeds extend over nearly 9,600 acres in central Colorado at elevations ranging from 2,500 to 2,700 m (8,200–9,000 ft) (McLeroy and Anderson, 1966). The National Monument consists of approximately 5,998 acres, protecting the southern arm of the Florissant Lake Beds.

Radiometric dates of sanidine crystals recovered from within the Florissant Formation and the underlying Wall Mountain Tuff and biostratigraphic age estimates indicate that the formation is latest Eocene in age (MacGinitie, 1953; Epis and Chapin, 1975; Evanoff et al., 2001). Samples from the three highest stratigraphic levels within the Florissant Formation produced a mean sanidine $^{40}Ar/^{39}Ar$ single-crystal age of 34.07 ± 0.10 Ma, and samples taken from the underlying Wall Mountain Tuff produced an age estimate of 36.07 Ma (Evanoff et al., 2001).

Plant fossils are preserved as impressions in each of the shale units, and *in situ* fossil stumps are present at the base of the lower mudstone. These specimens are recorded in the FLFO database and were recovered from more than 50 different localities. Unfortunately, the great majority of specimens are only attributed to "Florissant" and not any particular quarry. The lack of precise field data for most specimens hinders comparative studies among localities of the lake beds and determination of which fossil specimens were initially collected within park boundaries. MacGinitie (1953) mapped three historical localities – the Denver Museum of Natural History locality (collected in 1915 or 1916), the "Princeton" locality (collected in 1880), and Scudder's initial locality (collected in 1879) – as well as the localities University of Colorado 3731–3733, and tabulated abundances of specimens he recovered from four different quarries; he collected these specimens while conducting a revision of the Florissant flora (table 2 in MacGinitie, 1953). Each of these localities, except for the Denver Natural History Museum locality, appear to fall within FLFO.

The plant macrofossils of Florissant have been studied by numerous researchers over the past 150 years (e.g., Lesquereux, 1878; Cockerell, 1908; Knowlton, 1916; MacGinitie, 1953; Manchester, 2001; see Meyer, 2003, for an overview). The first comprehensive revision of the fossil flora was published by MacGinitie (1953). He recognized 115 species of bryophytes, horsetails, ferns, gymnosperms, and angiosperms in 84 genera and 44 families, and an additional 29 taxa *incertae sedis*, which included extinct species of uncertain taxonomic affinity and specimens too poorly preserved for accurate taxonomic assessment (MacGinitie, 1953). In an update to MacGinitie's monograph, which includes revisions of previously published species as well as 18 recently discovered taxa, Manchester (2001) reported 84 genera and 96 species of leaves, 70 genera and 73 species of reproductive structures (flowers, fruits, seeds), and five genera and six species of fossil wood. Thus, the flora contains as many as 169 species if it is assumed that there is no taxonomic overlap between leaves, wood, and reproductive structures assigned to the same genus but not yet found in organic connection. This approximation is roughly equal to the number of pollen types recognized in the flora (Leopold and Clay-Poole, 2001; Wingate and Nichols, 2001; Bouchal et al., 2016). Manchester (2001) also prepared a pared-down list of reliably identified seed plants that includes 46 genera in 28 families of angiosperms and six genera in three families of gymnosperms. Many of the most spectacular specimens are illustrated in full color in the book *Fossils of Florissant* (Meyer, 2003). A database with color images of type specimens of Florissant fossils acquired by Herb Meyer is accessible online (https://florissant database.colorado.edu).

Among angiosperms, the families with the greatest megafossil diversity at Florissant are the Fabaceae (3–9 genera, 5–11 spp.), Fagaceae (2 genera, 10 spp.), Rosaceae (9 genera, 11 spp.), Sapindaceae (3–8 genera, 7–11 spp.), and Ulmaceae (3 genera, 5 spp.). Fabaceae, Ulmaceae, and Sapindaceae are the only families to be represented by fossil leaves, wood, and reproductive structures (MacGinitie, 1953; Manchester, 2001; Wheeler, 2001). The most diverse genera are *Quercus* (9 spp.), *Rhus* (3 spp.), *Acer* (3 spp.), *Crataegus* (3 spp.), and *Salix* (3 spp.). The most abundantly represented in terms of specimens are the extinct genera *Cedrelospermum* (formerly assigned to *Zelkova* by MacGinitie, 1953) and *Fagopsis*, an extinct oak relative. Although not as abundant or diverse as the dicotyledonous component of the flora, several monocotyledonous taxa have been described from Florissant based on leaves, stems, and reproductive material (Manchester, 2001). The extant monocot genera *Potamogeton*, *Stipa*, and *Dioscorea* have been reliably identified, and *Typha* and *Smilax* have putative records (MacGinitie, 1953; Manchester, 2001). A single fan-palm leaf specimen was recovered (pl. 3, figure 1 in Manchester, 2001) but lacks sufficient detail to assign to a particular extant genus and was therefore assigned to the fossil genus *Palmacites*.

Among gymnosperms, the most diverse families are Pinaceae and Cupressaceae. Pinaceae are represented by seeds, cones, three-needled fascicles, and five-needled fascicles attributed to the extant genera *Abies*, *Picea*, and *Pinus*; Cupressaceae are represented by wood, seed cones, pollen cones, and branchlets assigned to the extant genera *Chamaecyparis* and *Sequoia* (MacGinitie, 1953; Manchester, 2001; Meyer et al., 2008).

Non-seed plants also form a minor component of the flora. All fern taxa historically described from Florissant are now treated as a single species, *Dryopteris guyottii* (Lesquereux, 1883; Cockerell, 1908; MacGinitie, 1953). Additional records of non-seed plants include the horsetail *Equisetum florissantense* and two species of *Plagiopodopsis* (Figure 13), an extinct genus in the bryophyte family Grimmiaceae (Cockerell, 1915; Britton and Hollick, 1907, 1915; Steere, 1947; MacGinitie, 1953).

In addition to the megafossil leaves, flowers, and fruits preserved as impressions in the shale, the Florissant Formation is also famous for petrified wood, including the iconic huge stumps of *Sequoia* near the Park visitor center. Angiosperm wood is also present but occurs in smaller pieces, many of which have probably been lost to years of private collecting prior to establishment of the park. Several wood types have been described (Wheeler, 2001; Wheeler and Meyer, 2012).

Detailed investigations of Florissant pollen and spores, both by light microscopy (Leopold and Clay-Poole, 2001; Wingate and Nichols, 2001), and by combined light and electron microscopy (Bouchal et al., 2016), provide more evidence in support of taxa known from megafossils, plus information on the presence of additional taxa not known from megafossils, such as *Cathaya* (Pinaceae), Onagraceae, *Tetracentron* (Trochodendraceae), and *Diospyros* (Ebenaceae).

The recovery of well-preserved reproductive fossils and revision of material originally assigned to extant genera has resulted in the recognition of a greater number of extinct genera at Florissant and undermines the assumption that Paleogene fossil leaves can be placed unequivocally in extant genera (Manchester and Crane, 1983, 1987; Manchester, 1989a, 1992, 2001; Manchester and Donoghue, 1995; Wang and Manchester, 2000). Seven extinct genera of angiosperms have been described from the Florissant flora: *Fagopsis* (Fagaceae; Figure 13), *Cedrelospermum* (Ulmaceae), *Asterocarpinus* (Betulaceae), *Diplodipelta* (Caprifoliaceae), *Florissantia* (Malvaceae), *Chaneya* (aff. Simaroubaceae), and *Palaeopotamogeton* (aff. Potamogetonaceae).

3.3.2 Greater Yellowstone Inventory and Monitoring Network
Yellowstone National Park

Yellowstone National Park (YELL), incorporated in 1872 as the first National Park in the nation, encompasses 2.2 million acres in northwestern Wyoming. The geology of this area includes a remarkably complete and fossiliferous stratigraphic record, preserving plants, vertebrates, invertebrates, and trace fossils in dozens of geologic units that span 540 million years. Significant paleobotanical remains are preserved in three formations of the Absaroka Volcanic Supergroup, which encompasses all Eocene volcanic and volcaniclastic rocks in the vicinity of Yellowstone (Smedes and Prostka, 1972), and in Quaternary lake sediments.

Lamar River Formation: The Lamar River Formation represents volcaniclastic Lower Eocene (ca. 51–49 Ma) sediments in the Lamar River Valley (Fritz, 1982). This unit contains two of Yellowstone's best known fossil forests at Amethyst Mountain and Specimen Ridge (Knowlton, 1899; Dorf, 1964; Fritz, 1982; Wheeler, 2024), as well as numerous horizons that have yielded fossil leaves (Figure 13; Lesquereux, 1872; Knowlton, 1899; Dorf, 1964).

The fossil stumps and logs preserved at Specimen Ridge were initially interpreted as representing as many as 27 discrete fossil forests buried *in situ* by episodic deposition of volcanic debris (Dorf, 1964). Subsequently, observations of high-energy mudflows transporting stumps from high to low elevations in modern volcanic regions, stratigraphic cooccurrence of fossil taxa whose modern relatives have nonoverlapping climatic tolerances, and fossil woods with and without seasonal growth rings inspired extensive discussion of the extent of taphonomic mixing that occurred during deposition and preservation of the Specimen Ridge flora (Dorf, 1960, 1964; Fritz, 1980, 1982, 1986; Yuretich, 1984; Wing, 1987; Wiemann et al., 1999).

The material that buried these forests consists of conglomerates, breccias, and tuffaceous siltstones and sandstones; fossil wood is associated with the former two lithologies, as well as paleosols, and fossil leaves are found as compressions and impressions in the tuffaceous material (Knowlton, 1899; Dorf, 1964; Fritz, 1982). The wood assemblage at Specimen Ridge is numerically dominated by the conifer families Cupressaceae and Pinaceae but also includes several angiosperm families, notably Anacardiaceae, Betulaceae, Cornaceae, Fagaceae, Juglandaceae, Lauraceae, and Ulmaceae (Knowlton, 1899; Beyer, 1954; Wheeler et al., 1977, 1978; Wheeler, 2024).

In contrast to the wood assemblages, angiosperms dominate the leaf floras described by Knowlton (1899), both numerically and in diversity. Knowlton described more than 139 species of fossil leaves from the leaf flora, though this is likely an overestimation of the number of species represented by the fossils. Two of the authors of this contribution (Huegele and Manchester) examined Knowlton's original published diagrams and photographs of his original specimens housed at the Smithsonian National Museum of Natural History in Washington, DC, and observed that, although many taxa seem correctly identified, many others need revision or synonymy (see Supplementary Table 8: Preliminary Review of Leaf Taxa). The impression flora contains horsetails (*Equisetum* stems) and fern foliage (Figure 13) reviewed by Collinson (2001, 2002), several conifers, and numerous angiosperm plant families, including all of the families represented by wood in addition to Fabaceae (Knowlton, 1899; Dorf, 1964; Herendeen et al., 2022). It is also possible that Musaceae, Ebenaceae (Figure 13), Cercidiphyllaceae, Sabiaceae, Betulaceae, Cornaceae, Moraceae, Lauraceae, and Magnoliaceae are represented in the leaf flora, but detailed study will be needed to confirm their presence (Figure 13; Supplementary Table 8). Some potentially notable taxa among the leaf flora include tropical elements like *Musaphyllum* (foliage of Musaceae) and *Artocarpus*, whose presence, if confirmed, would imply warm and wet environmental conditions based on the current distributions of their living relatives. The need for systematic revision of this flora has been alluded to by previous authors (Andrews and Lenz, 1946) and should be considered a resource management priority for paleontology in YELL.

The fossil forests of Amethyst Mountain were first described more than 100 years ago (Holmes, 1879; Platen, 1908). As at Specimen Ridge, petrified wood is preserved in a sequence of vertically stacked layers that represent forests preserved in distinct depositional episodes (Wheeler et al., 1977, 1978; Fritz, 1980, 1982; Wheeler, 2024). A detailed study of 64 stumps and logs from a single lithological unit in this sequence revealed that upright stumps, stumps that were upright but offset from vertical, and horizontal logs occurred together in a single layer, and this cooccurrence likely indicates a mixture of *in situ* and transported material (Fritz and Fisk, 1978). Preliminary identifications of the wood from this layer provide similar results to detailed systematic studies from Specimen Ridge. Many of the same families are present, and angiosperm wood types represent a large portion of taxa but a small percentage of individuals (Wheeler et al., 1977, 1978; Fritz and Fisk, 1978; Wheeler, 2024). Exposures of the Lamar River Formation at Amethyst Mountain also preserve fossil leaves described by Lesquereux (1872) and Knowlton (1899).

Sepulcher Formation: The Sepulcher Formation is a lateral equivalent of the Lamar Ridge Formation exposed in the northwest region of the park and in the geographically adjacent Gallatin National Forest (Fritz, 1982). Fossil forests like those preserved in the Lamar River Formation at Specimen Ridge and Amethyst Mountain have been identified within YELL at Specimen Creek and Bighorn Peak (Andrews, 1939; Andrews and Lenz, 1946; Wheeler et al., 1977, 1978; Chadwick and Yamamoto, 1984; Wheeler, 2024). Nine wood-bearing stratigraphic levels were identified at Specimen Creek, and each preserves a mixture of upright stumps and horizontal logs (Chadwick and Yamamoto, 1984). Taxonomic groups present at Specimen Creek but not known from the wood floras of the Lamar River Formation include Ebenaceae, Hamamelidaceae, and Salicaceae (Chadwick and Yamamoto, 1984). The Sepulcher Formation also includes some of the source localities for the leaf and fruit fossils described by Knowlton (1899).

Langford Formation: The Middle Eocene (ca. 46–47 Ma) Langford Formation is extensively exposed in the Lamar River Valley, where it unconformably overlies the Lamar River Formation (Fritz, 1994; Santucci, 1998). A paleobotanically fossiliferous horizon of the Langford Formation was discovered in a road cut between Lake Butte and Sylvan Pass during construction of the East Entrance Road (Fritz, 1994; Santucci, 1998). This horizon produced many large leaves of an extinct genus of sycamore, *Macginitiea* (Fritz, 1994), which are now held in the Yellowstone Park collection.

Quaternary sediments: Yellowstone National Park includes several sections that preserve late Pleistocene (ca. 140,000 ybp to present day) palynomorphs and plant macrofossils (Baker, 1976, 1986; Baker and Richmond, 1978). The southeast arm of Yellowstone Lake has been particularly productive, yielding palynomorphs in lake cores and plant macrofossils in stream cuts along tributaries to Yellowstone Lake (Baker, 1970, 1986). The palynomorph assemblages are numerically dominated by *Pinus* and *Picea*, but also include other conifers, *Abies* and *Pseudotsuga*, and the dicot trees *Alnus, Betula, Quercus*, and *Salix*. Macrofossil assemblages record many of these same taxa, in addition to herbs such as *Potentilla, Gaultheria*, and *Selaginella*. The combined sequence of pollen and macrofossils throughout the Sangamonian and Wisconsin glacial–interglacial cycles suggests that vegetation transitioned from dense, coniferous forests to open, parkland, coniferous forests or tundra during cool periods (Baker, 1986).

Grand Teton National Park

Grand Teton National Park (GRTE) preserves part of the Teton Range of northwestern Wyoming. Rocks exposed within the park range from Cambrian to Quaternary and contain minor Cenozoic paleobotanical material in the middle Eocene Hominy Peaks Formation and the Pliocene Teewinot Formation.

Hominy Peaks Formation: The Hominy Peaks Formation is made up of volcaniclastic rocks and encompasses all Eocene exposures within the Teton Range (Leopold et al., 2007). Exposures of the formation near the northwest corner of GRTE preserve fragments of carbonized plant remains, mats of contorted plant debris, and fossil pollen at several stratigraphic horizons. Palynomorphs from the Hominy Peaks Formation include eight types of ferns, eight types of gymnosperms, 26 types of eudicots, and three types of monocots. Biostratigraphic correlation of this assemblage was used to establish a Middle Eocene age, and its taxonomic composition resembles that of the Middle Eocene Kisinger Lakes flora of northwestern Wyoming (MacGinitie, 1974; Leopold et al., 2007). East of GRTE, fossil-leaf–bearing horizons have been identified in the underlying Pinyon Conglomerate at Whetstone Mountain and in unnamed Eocene rocks exposed at Togwotee Pass (Tweet et al., 2013)

Neogene Resources: Minor occurrences of paleobotanical material have been reported from several Neogene units within GRTE. Pliocene palynomorphs have been recovered from the Shooting Iron and Teewinot formations (Leopold et al., 2007), and coniferous wood and a possible pinecone fragment were identified from the Teewinot Formation (Tweet et al., 2013). In Quaternary sediments, Santucci and Wall (1999) reported a piece of ca. 27 kya carbonized wood found in a laminated silt near Pilgrim Creek, and Holocene charcoal recovered from the area around Jackson Hole (including within GRTE) and palynomorphs recovered from Hedrick Pond in GRTE provide a record of recent fire and vegetation history (Pierce and Good, 1992; Jacobs and Whitlock, 2006, 2008).

3.3.3 Northern Colorado Plateau Inventory and Monitoring Network

Fossil Butte National Monument

Fossil Butte National Monument (FOBU), just west of the town of Kemmerer in southwestern Wyoming, has attracted interest since the late nineteenth century because of the wealth of fossil fish and other vertebrate remains preserved in fissile lacustrine shales of the Green River Formation (Fossil Butte Member). Green River Formation fossils continue to be popular for sales to museums, interior decorators, and fossil enthusiasts. Although amateur collecting of these fossils is now prohibited on Bureau of Land Management and National

Monument land, major discoveries continue to be made in privately owned or leased fish quarries outside of the monument. Fossil plant remains are recovered mainly as a byproduct of fish excavation, but the iconic large palm fronds are highly sought-after and expensive. Other plant remains are diverse, but well-preserved specimens are relatively rare.

Green River Formation: The prolific shales occur in the Fossil Butte Member of the Green River Formation and date to about 52 Ma (Grande, 2013) based on radiometric analyses. This member is older than the Parachute Creek Member of the same formation, which is well known for fossil plants recovered from numerous sites in eastern Utah near Bonanza and western Colorado north of Rifle and in the area of Douglas Pass (MacGinitie, 1969).

The plant fossils recovered from FOBU are impressions in shale recovered by splitting with thin, wide chisels. The flora of Fossil Butte is too diverse to list exhaustively, encompassing more than 250 leaf types and 100 fruit types (Tweet et al., 2012c). Here we call attention to some of the notable remains collected from the monument. The leaves include aquatic plants such as *Salvinia*, *Ceratophyllum*, and *Nelumbo*), as well as a wide range of terrestrial plants. Distinctive winged fruits of *Ailanthus* are among the more common plant remains. *Platycarya*, a genus of the walnut family now confined to eastern Asia, is commonly found, represented by leaflets and catkins (Wing and Hickey, 1984). The extinct elm-related plant *Cedrelospermum* is represented both by leaves and winged fruits. However, many of the other constituents have not been satisfactorily identified and some probably represent as-yet-undescribed extinct genera. The extinct rabbit ear-winged fruit, *Lagokarpos*, is relatively common in these beds and is shared with other Eocene lacustrine strata in North America, Germany, and Tibet (McMurran and Manchester, 2010; Tang et al., 2019). A fortunate aspect of these shales is that they can be readily collected in large slabs or sheets, so there is a greater chance of recovering more complete and informative specimens. In rare instances, leaves and fruits are attached to the same twig, providing direct proof that these organs represent the same species of plant. Several plant taxa were described from this flora by Brown (1937), but the identities of some, such as those he attributed to *Ficus*, remain uncertain. The most comprehensive treatment of the flora, illustrating the best examples of many different morphotypes, was published in color by Lance Grande in his book the *Lost World of Fossil Lake* (Grande, 2013).

Cedar Breaks National Monument

Cedar Breaks National Monument (CEBR) is located in southwest Utah, approximately 90 km (56 mi) west of Bryce Canyon National Park, and was established to preserve the Cedar Breaks amphitheater (Tweet et al., 2012c). The monument has similar geology to BRCA (see below) and also preserves endocarps belonging to either *Celtis* or *Prunus* in the Paleocene–Middle Eocene Claron Formation (Schneider, 1967). Other paleobotanical material recovered from CEBR includes pollen, conifer needles, and wood of *Picea* found in a core from Alpine Pond that spans the last 3,000 years (Anderson et al., 1999). Other Quaternary sites close to the boundaries of the park preserve pollen, conifer needles, and other plant macrofossils from the Late Pleistocene (Mulvey et al., 1984; Anderson et al., 1999; Madsen et al., 2002).

Bryce Canyon National Park

Bryce Canyon National Park (BRCA) preserves approximately 35,835 acres in southwest Utah and is known for its scenic geologic features, including pinnacles, spires, and horseshoe-shaped amphitheaters. Paleobotanical resources contained within BRCA are limited to some discoveries within the Paleocene–Middle Eocene Claron Formation (Tweet et al., 2012c). Seeds belonging to either *Celtis* or *Prunus* have been reported from the Claron Formation, and the source layer has been correlated with a fossiliferous layer of the Claron Formation in Cedar Breaks National Monument, roughly 90 km (about 56 mi) west of BRCA (Schneider, 1967). More recently, *Celtis* endocarps and leaf impressions (Figure 13) were collected and identified in 2013 (Tut Tran, NPS, pers. comm., 2023). Leaf impressions have also been reported (Gregory, 1950), but none were identified or collected. Other paleobotanical material from the Claron Formation of BRCA includes molds of angiosperm seeds and fossilized wood in the park collections (Tut Tran, NPS, pers. comm., 2023), and root traces preserved in the formation's upper member (Taylor, 1993; Tweet et al., 2012c). Age constraints on the Claron Formation are provided by a Middle Eocene rodent jaw collected from its upper member, and radiometric dating of volcaniclastic strata in the overlying Brian Head Formation to 36 Ma (Biek et al., 2011; Eaton et al., 2011).

3.3.4 Southern Colorado Plateau Inventory and Monitoring Network

Aztec Ruins National Monument

Aztec Ruins National Monument (AZRU) preserves the ruins of a twelfth-century Pueblo Indian community in northwestern New Mexico. The bedrock of this unit is composed of the fossiliferous Nacimiento Formation, but the unit only occupies a small area (318 acres) with few geologic exposures (KellerLynn, 2016). Paleobotanical material found within AZRU is sparse and includes unidentified petrified wood, a possible leaf, and a possible seed held in the park collections. Petrified wood modified by humans is found in artifact assemblages and reworked petrified wood is found in Quaternary terrace deposits (Tweet et al., 2009; KellerLynn, 2016). The original source of this wood is unknown, but the Paleocene Nacimiento Formation, which contains petrified wood and fossil leaves outside of AZRU and is mapped nearby, is a likely candidate.

El Malpais National Monument

El Malpais National Monument (ELMA) in western New Mexico is a minor source of paleobotanical resources. Tree molds preserved in lava flows can be seen in the Bandera flows (11,000 ybp) at Ke Cave Trading Post (Laughlin and Perry, 1997). Charcoal and "baked" roots from the Lava Crater (17,000 ybp), Bandera, and McCartys flows (3,000 ybp) have been radiocarbon dated (Laughlin et al., 1993, 1994; Mabery, 1997). Quaternary-aged sub-fossil wood was also found and studied in ELMA for climate data (Grissino Mayer and Swetnam, 2000; Poore et al., 2005).

3.3.5 Sonoran Desert Inventory and Monitoring Network

Tuzigoot National Monument

Tuzigoot National Monument (TUZI) protects the remains of a pueblo built by the Sinagua people between about 1100 and 1450 and a surrounding area of about 800 acres. It is located about 30 km (19 mi) northwest of Montezuma Castle and has similar geology to that unit (see below). Undescribed plant fossils from TUZI have been found at a locality established by the Museum of Northern Arizona (MNA 330–0) in Pliocene strata of the Verde Formation (Tweet et al., 2008).

Montezuma Castle National Monument

Montezuma Castle National Monument in central Arizona preserves cliff dwellings built by the Sinagua people in the twelfth and thirteenth centuries and a large permanent spring, Montezuma Well. Cenozoic paleobotanical material has been found from a few Neogene and Quaternary sites in the Montezuma Well unit. The Miocene–Pliocene Verde Formation preserves plant fossils at a locality established by the Museum of Northern Arizona (MNA 295–0) near Montezuma Well and in the limestone walls of the well (Lange, 1957; Tweet et al., 2008). The fossils in the well walls are reported as "reed-like" impressions and molds of plant debris (Lange, 1957). Pleistocene plant impressions and palynofloras have been identified in travertine deposits dated to between 400,000 and 700,000 ybp nearby the well. The macrofloral impressions include leaves of trees and riparian shrubs and a pinecone similar to extant *P. ponderosa* (Hevly et al., 1992). Cores taken from the permanent spring at the bottom of Montezuma Well have yielded pollen and plant macrofossils, which have been used to reconstruct the site's Quaternary environmental history (Hevly, 1974; Davis and Shafer, 1992, 2006; Hevly et al., 1992; KellerLynn, 2019).

3.3.6 Southern Plains Inventory and Monitoring Network

Alibates Flint Quarries National Monument and Lake Meredith National Recreation Area

Alibates Flint Quarries National Monument and Lake Meredith National Recreation Area (LAMR) in western Texas are adjoining regions managed jointly by the NPS that share the same geology. The most notable paleobotanical material associated with the two units comes from the Miocene–Pliocene Ogallala Formation. Wilson (1988) reported root casts, silicified grass anthoecia, *Celtis* endocarps, and a Clarendonian (13.6–10.3 Ma) vertebrate assemblage from the informal "LX" member of the Ogallala Formation. However, it is unclear what specimens were found within which NPS unit (Hunt and Santucci, 2001). Other paleobotanical resources include plant debris in an ash bed known as the Lava Creek ash (about 640 kya) in LAMR's South Canyon, and likely reworked petrified wood alongside proboscidean remains in Pleistocene gravels (Hunt and Santucci, 2001; Tweet et al., 2015).

3.3.7 Chihuahuan Desert Inventory and Monitoring Network

Old Spanish National Historical Trail

The Old Spanish National Historic Trail (OLSP) protects a network of Indigenous trails, early trade and exploration routes, and horse and mule paths that span approximately 4,300 km (2,700 mi) and cross California, Nevada, Arizona, Utah, Colorado, and New Mexico. Petrified wood was found along OLSP in Cañon Largo, New Mexico, during the 1859 Macomb Expedition, a government-funded expedition with the goal of developing a route to transport military supplies between Santa Fe, New Mexico, and southern Utah (Madsen, 2010). The wood is unidentified but is likely derived from the Paleocene Nacimiento Formation or the Eocene San Jose Formation. The potential for recovery of Cenozoic paleobotanical material along much of OLSP is limited (the youngest bedrock predates the Cenozoic), with a few exceptions. The trail passes through or nearby several other NPS-managed areas known to contain Cenozoic paleobotanical resources, including MOJA, LAKE, and AZRU.

White Sands National Park

White Sands National Park (WHSA) in south-central New Mexico preserves one of the world's largest gypsum and sand dune fields. Cenozoic plant macrofossil remains from WHSA are present in Quaternary sediments and include external molds of cacti and other plants that became encrusted with or replaced by gypsum crystals (Vandiver, 1936; Santucci et al., 2007), and layers of *Ruppia cirrhosa* (ditchgrass) endocarps (Bennett et al., 2021a). The ditchgrass endocarps are preserved at several stratigraphic horizons within a section that also includes palynomorphs and tracks of humans, proboscideans, and canids. Radiocarbon dating of the endocarps and palynomorphs constrain the age of seven stratigraphic horizons preserving human tracks to approximately 23–21 kya within the Last Glacial Maximum (Bennett et al., 2021a, 2021b; Pigati et al., 2023). These tracks provide a temporal extension of the coexistence of early human inhabitants of North America and Pleistocene megafauna.

Big Bend National Park

Big Bend National Park (BIBE) in southwestern Texas conserves over a million acres of Chihuahuan Desert and is very well known for its Cretaceous and Early Cenozoic assemblages of fossil plants and vertebrates. The Cenozoic paleobotanical resources of BIBE include petrified wood, leaves, and root traces that occur in four formations that span the Early Paleocene to Early Oligocene. Geographically, these resources are concentrated in the Tornillo Flats of the northeast-central section of the park where the Tornillo Group, defined by the Javelina, Black Peaks, and Hannold Hill formations, crops out extensively.

Black Peaks Formation: The Blacks Peaks Formation is stratigraphically positioned between the Javelina and Hannold Hill formations, and collectively these units constitute the Tornillo Group, a succession of Upper Cretaceous, Paleocene, and Lower Eocene fluvial strata that preserve abundant floral and faunal remains (Lehman et al., 2018). Two intervals in the Black Peaks Formation contain well-preserved petrified wood. The "log-jam" level is a thin but widespread stream-channel sandstone that preserves many thousands of logs and is dated to the Early Paleocene (Torrejonian; 63.3–60.2 Ma) based on mammal biostratigraphy (Wheeler and Lehman, 2009). The wood recovered from this horizon includes three angiosperm taxa, *Paraphyllanthoxylon abbottii* (affinity to Burseraceae), cf. *Plataninium haydenii* (affinity to Platanaceae), and cf. *Cunonioxylon sensu* Gottwald (affinity to Cunoniaceae), but most specimens (all but two) studied belong to the taxon *P. abbottii* (Wheeler, 1991; Wheeler and Lehman, 2009). Two types of araucarioid wood (BP 8 and 9 of Wheeler and

Lehman, 2005) have been identified from higher levels in the formation, above an erosional surface overlain by Eocene and younger strata. Exact locality information for these specimens is on file at the North Carolina Museum of Natural Sciences (Wheeler and Lehman, 2009).

Hannold Hill Formation: The overlying Hannold Hill Formation is exposed in the Tornillo Flats Region of BIBE and comprises mudstones, sandstones, and conglomerate deposited in an inland floodplain setting (Wick and Corrick, 2015). Some petrified wood is found as rounded or reworked clasts within a sandstone conglomerate (Lehman et al., 2018). Early Eocene (Wasatchian; 55.4–50.3 Ma) vertebrate assemblages have been recovered from sandstones in the upper part of the Black Peaks Formation and lower part of the Hannold Hill Formation (Lehman et al., 2018).

Canoe Formation: The Middle Eocene Canoe Formation unconformably overlies the Hannold Hill Formation and comprises conglomerate, sandstones, siltstones, and mudstones deposited in braided stream and floodplain settings (Wick and Corrick, 2015; Wick, 2021). Paleobotanical resources of the Canoe Formation include leaves, reproductive structures, and transported, *in situ*, and reworked fossil wood (Figure 13). The Big Yellow Sandstone Member contains Eocene conifer and angiosperm wood, abraded fragments of wood believed to have been exhumed from underlying Cretaceous strata during Eocene entrenchment, and, in the McKinley Hills area of BIBE, a small forest of 92 tree stumps (Wick, 2021). These stumps have been described as poorly preserved and the samples taken were not viable for analysis or identification, but the site is extensive and has potential for further study (Tom Lehman, pers. comm., 2023). Limonite-encrusted logs are also found within the park (Lehman et al., 2018). Fossil leaves and reproductive structures are preserved in siltstone and lignite layers originally attributed to the Javelina Formation (Lawson, 1972), but subsequent stratigraphic study indicates that the strata and fossils represent the Canoe Formation (Turner et al., 2011; Wick, 2021). Based on the fossil leaf impressions, Lawson (1972) identified six taxa: *Ampelopis acerifolia*, *Ficus* cf. *F. tennesseensis*, *Laurus socialis*, *P. raynoldsi*, *Saccoloma gardneri*, and *Typha*.

Chisos Formation: Overlying the Canoe Formation, the Chisos Formation shows evidence of past volcanism and includes igneous intrusions, lavas, ashfalls, and fluvial volcaniclastic sediments deposited between ca. 42 and 32 Ma (Wick and Corrick, 2015). The Chisos Formation has been reported to contain petrified wood, although occurrences are rare, and no specimens have been identified (Maxwell et al., 1967). There is also a record of root traces being common in certain volcanic areas (Wilson and Runkel, 1989).

El Camino Real de los Tejas National Historical Trail

El Camino Real de los Tejas National Historical Trail (ELTE) designates approximately 4,000 km (2,500 mi) of Indigenous trails and trade routes that connected Mexico City and Los Adaes (the capital of Spanish Texas prior to 1770), and were heavily used by Spanish colonists during their occupation of land that is today Texas and northwest Louisiana (Figure 10). Cenozoic paleobotanical resources are present along certain parts of the trail. Stone City Bluff is a fossil site on ELTE that has produced unidentified plant debris (Stenzel et al., 1957; Hodgkinson, 1974). A Quaternary-aged mammoth site on the trail also includes plant debris and pollen (Carpenter et al., 2013). In some parts of Texas and Louisiana, the trail roughly follows sediments belonging to the Paleocene and Eocene Wilcox and Claiborne groups that have produced diverse fossil floras in Texas and Tennessee, respectively. Exposures of the Wilcox Group within a mile of the trail in this region have yielded abundant fossil leaves (Deussen, 1924). There is also potential to discover fossil wood from exposures of the Cenozoic Catahoula and Yegua formations located along the trail in south-central Texas.

3.4 Midwest Region

The Midwest Region includes 53 NPS units in Illinois, Indiana, Iowa, Kansas, Michigan, Minnesota, Missouri, Nebraska, North Dakota, Ohio, South Dakota, and Wisconsin (Figure 14). These units are divided into three inventory and monitoring networks: the Great Lakes Network, the Heartland Network, and the Northern Great Plains Network (NGPN). Collectively, they preserve paleobotanical resources in 11 different NPS units and 18 different formations that span the Paleocene to Holocene.

The NPS units of the Midwest Region preserve significant paleobotanical resources in the Paleocene petrified forests of Theodore Roosevelt National Park and within Late Eocene–Late Miocene deposits of South Dakota and Nebraska units that are thought to document the early development of the North American grassland ecosystems that characterize the natural vegetation of much of the Midwest today (Figures 15 and 16; Fastovsky and McSweeney, 1991; Strömberg and McInerney, 2011). The assemblage of *in situ* stumps, fossil leaves, and reproductive structures preserved in the Upper Paleocene Sentinel Butte Formation at the Petrified Forest Plateau of Theodore Roosevelt National Park offers insight into the floristic composition and paleoecology of mid-latitude Paleocene forests and a point of comparison with the well-documented localities of the Sentinel Butte Formation outside of the park (Crane et al., 1990; Manchester et al., 2004) and with the Middle and Late

Figure 14 Map of NPS units with Cenozoic paleobotanical resources in the Midwest NPS, colored by inventory and monitoring network. See Figure 1 legend for details on the park symbology and acronyms.

Park Unit	Significance	PA	E	O	M	PL	Q
FOUS							
THRO	Highest	PA	E				
WICA				O			Q
BADL			E	O			
AGFO					M		
SCBL				O			
CHRO							
NIOB					M		
PIRO							Q
ICAG							
HEHO							Q

Confirmed / Unconfirmed
Highest / High / Moderate

Figure 15 Summary of the temporal distribution and significance of Cenozoic paleobotanical resources in the Midwest Region. See Figure 2 legend for explanation of significance ratings

Epoch	Northern Great Plains				
	THRO	BADL	SCBL	AGFO	NIOB
Miocene					Ash Hollow
					Valentine
				Anderson Ranch	
			Upper Sharps	Harrison and Anderson Ranch, undivided	Harrison
Oligocene			Lower Sharps	Gering	
			Brule	Brule	
			Chadron		
			Chamberlain pass		
Eocene					
Paleocene		Golden Valley			
		Sentinel Butte			
		Bullion Creek			

Figure 16 Stratigraphic chart of Tertiary plant-fossil–bearing formations in the Midwest Region.

Paleocene assemblages of the Fort Union Group in Wyoming and Montana (Wing et al., 1995; Wilf, 2000; Peppe, 2010), among other North American Paleocene floras (see Wing, 1998; Graham, 1999; and Pigg and DeVore, 2010 for reviews). Collectively these Paleocene floras document the forested ecosystems that emerged several million years after the K–Pg mass extinction and provide a regional context for interpreting episodes of pulsed taxonomic turnover at the Paleocene–Eocene boundary documented from individual basins (Wilf, 2000; Wing et al., 2005).

The history of North American grasslands has been inferred using four lines of (paleo)botanical evidence: macrofossils, microfossils (pollen and phytoliths), paleosols, and molecular phylogenies (Graham, 1999; Edwards et al., 2010; Strömberg, 2011). Three of these types of evidence (macrofossils, microfossils, and paleosols) are preserved within or adjacent to NPS units of the Midwest Region. Paleosols described from Lower Oligocene strata of the White River Group in Badlands National Park (Retallack, 1983) and Lower Miocene strata of the Arikaree Group in Agate Fossil Beds National Monument (Retallack, 1997), phytoliths collected from exposures of Brule Formation and undivided Miocene Monroe Creek and Harrison Formations with correlative strata in Scotts Bluff National Monument (Strömberg, 2004), and grass macrofossils collected from the Ash Hollow Formation with correlative strata in Niobrara National Scenic River (Thomasson, 1980) support three different scenarios regarding the age of open grassland ecosystems in the Great Plains (reviewed in Strömberg, 2011).

3.4.1 Northern Great Plains Inventory and Monitoring Network

Fort Union Trading Post National Historic Site

The Fort Union Trading Post National Historic Site (FOUS) is located on the North Dakota–Montana state line south of Williston, North Dakota. Within FOUS, small pieces of petrified wood are commonly found amongst gravels that have been reworked by fluvial and glacial transport and are likely derived from the Paleocene Fort Union Group (Tweet et al., 2011; Tweet, 2016).

Theodore Roosevelt National Park

Theodore Roosevelt National Park (THRO) preserves about 70,000 acres of badlands in North Dakota. The park contains thick exposures of Paleocene rocks that represent three different formations well-known for their preservation of fossil wood, leaves, and reproductive structures at localities outside the park (Hickey, 1977; Crane et al., 1990; Erickson, 1999). Recently, park monitoring surveys have discovered similar material within THRO (Salcido et al., 2022) and suggest potential for further paleobotanical investigation.

Bullion Creek Formation: The Bullion Creek Formation records fluvial, lacustrine, and estuarine deposition from ca. 61 to 59 Ma (Hoganson and Murphy, 2003; Belt et al., 2004; Anderson et al., 2006). Bullion Creek deposits in western North Dakota have a characteristic sequence of gray claystone and siltstone, lignite, yellow siltstones and sandstones, and sandstone (Salcido et al., 2022). These lithologies are interpreted as representing a succession of floodplain, swamp, crevasse splay, and point bar deposits. The formation preserves paleobotanical material, including leaves of *Cercidiphyllum* and unidentified petrified wood and seeds (Jacob, 1973; KellerLynn, 2007; Tweet et al., 2011).

Figure 17 Fossil plants from the Midwest Region. (a) Unidentified fossil leaf collected from the Sentinel Butte Formation in the North Unit of Theodore Roosevelt National Park. Photos by Charles Salcido. Scale bar = 1 cm. (b) *In situ* fossil log in the Sentinel Butte Formation of Theodore Roosevelt National Park. (c) Petrified wood found along the banks of the Niobrara River in Niobrara National Scenic River.

Outside of THRO, at Wannagan Creek Quarry, the Bullion Creek Formation preserves a diverse floral assemblage of more than 100 plant taxa and associated arthropod trace fossils based on leaf and fruit impressions in lignitic shale (Erickson, 1999; Haire et al., 2001).

Sentinel Butte Formation: The lithology of the Upper Paleocene Sentinel Butte Formation is similar to the underlying Bullion Creek Formation, but the Sentinel Butte Formation can be distinguished by its weathered color (gray vs. yellow) and abundance of petrified wood (Tweet et al., 2011). One particularly notable fossil locality within THRO is the Petrified Forest Plateau in the western section of the South Unit; this locality preserves fossil leaves and tree trunks (Figure 17), some of which are in their original growth position (Fastovsky and McSweeney, 1991).

The Petrified Forest Plateau preserves fossil trees in two organic-rich horizons, an upper lignite and a lower silicified layer, separated by 37 cm (15 in) of iron oxide–stained mudstones (Fastovsky and McSweeney, 1991). Most wood appears to be cupressaceous (formerly described as taxodiaceous), according to Fastovsky and McSweeney (1991, citing the help of Steve Manchester). Other wood includes *Dacrydium* and other podocarpaceous wood (unpublished reports by Coffin, 1984, and unpublished reports by his students, M. Arct and J. H Coulton) and *Sciadopitys* (Sciadopityaceae, formerly classified as Pinaceae; unpublished reports by B. F. Kukachka). Rare angiosperm wood has also been recovered, including a sample with similarities to *Cercidiphyllum* (Cercidiphyllaceae; Fastovsky and McSweeney, 1991, with the help of S. Manchester). The Petrified Forest Plateau also preserves leaves of *Cercidiphyllum* and *Platanus* and *Equisetum* stems in lenticular deposits of fine-grained sandstone (Fastovsky and McSweeney, 1991). Collectively, the presence of thick organic horizons in poorly developed paleosols, abundant upright tree trunks with buttressed bases and taxonomic affinity to Cupressaceae, and fragmentary mollusk and gastropod fossils suggest that these deposits accumulated in a lowland swampy area of a floodplain, potentially analogous to modern bald cypress swamps.

The Sentinel Butte Formation has produced additional paleobotanical material from other localities within and near the park. The South Unit Loop Survey (Salcido et al., 2022) reported the fruits of *Jenkinsella*, the revised name of Cercidiphyllaceae fruits formerly known as *Nyssidium* (Golovneva and Alekseev, 2017), and *Porosia* (fam. indet.), and Jain and Hall (1969) described the holotype of the fern megaspore *Azolla stanleyi* (Salviniaceae) from a horizon in the South Unit of THRO. Adjacent to the South Unit of THRO in Medora, North Dakota, silicified peat deposits preserve fossil leaves and wood of *Ginkgo*, *Metasequoia*, other conifers, and angiosperms (Ting, 1972; Harr and Ting, 1977), and quarries at the Medora Public Dig Site within the park preserve plant debris in carbonaceous mudstone.

Outside of THRO, the flora of the Sentinel Butte Formation is known from the Almont locality in Morton County and the Beicegel Creek locality in McKenzie County, North Dakota (Crane et al., 1990; Manchester et al., 2004; Pigg and DeVore, 2010). The Almont site preserves a diverse megaflora of 35–40 plant taxa based on leaf, fruit, and flower impressions and a palynoflora of more than 50 taxa (Crane et al., 1990; Pigg et al., 2008; Zetter et al., 2011). The Beicegel Creek site preserves many of the same taxa as Almont but in a slightly higher energy depositional environment and includes fossil fruits with organic preservation amenable to study through the cellulose acetate peel technique (Pigg et al., 2023).

Golden Valley Formation: The Golden Valley Formation conformably overlies the Sentinel Butte Formation and is divided into the lower Bear Den Member and the upper Camels Butte Member by laterally equivalent beds referred to as either the Taylor Bed or Alamo Bluff Lignite (Hickey, 1977; Tweet et al., 2011). These beds are used as markers for the Paleocene–Eocene boundary, ca. 56 Ma (Cohen et al., 2013). Within THRO, the Golden Valley Formation is only represented by erosional remnants of the Taylor Bed, a strongly resistant layer of silicified siltstone and freshwater limestone; the underlying Bear Den Member, composed mainly of siltstone and claystone, is thought to have eroded away (Hoganson and Campbell, 1997; Tweet et al., 2011). The Taylor Bed has produced plant fragments (unidentifiable roots, stems, and twigs; Hoganson and Campbell, 1997; KellerLynn, 2007), and root and stem casts (Biek and Gonzalez, 2001). Outside of THRO, the contemporaneous Alamo Bluff Lignite and slightly older Bear Den Member have yielded palynomorphs and a moderately diverse macrofossil assemblage of *Equisetum*, ferns, conifers, and monocot and dicot angiosperms (Hickey, 1977).

Unnamed gravel deposits: Unnamed terrace gravel deposits containing petrified wood are present in the North and South Units of THRO. The age of these deposits is uncertain but is broadly considered to represent Neogene deposition prior to Pleistocene glaciations (Tweet et al., 2011). It is also uncertain whether the petrified wood found in these deposits is reworked from older units (Salcido et al., 2022).

Wind Cave National Park

Wind Cave National Park (WICA) in southwestern South Dakota comprises about 34,000 acres of land surrounding Wind Cave, one of the most extensive cave systems in the world with more than 260 km (160 mi) of surveyed passages (Herring et al., 2023). The stratigraphy and paleobotany of Cenozoic strata within WICA is similar to that of other parks with the NGPN, and the park's primary Cenozoic paleobotanical resources are rhizoliths and *Celtis* endocarps from the Eocene–Oligocene White River Group. Such material has been recovered alongside invertebrate and vertebrate assemblages from several localities in the northeastern part of WICA, the Centennial Site and the Klukas Sites (Martin, 1984, 1988; Brown, 2014). Quaternary sediments of the Red Valley in southwestern WICA have produced phytoliths and charcoal (Fredlund and Tieszen, 1997).

Badlands National Park

Badlands National Park (BADL) in central South Dakota preserves approximately 240,000 acres and contains rich fossil assemblages in its Upper Cretaceous, Eocene, and Oligocene strata. Fossiliferous paleosols, sandstones, siltstones, and mudstones deposited from the Middle Eocene into the Early Oligocene belong to the White River Group. Although best known for their abundant vertebrate fossils, these strata also preserve significant paleobotanical remains in the form of wood, endocarps, and root traces (Tweet et al., 2011).

Chamberlain Pass Formation: The basal formation of the White River Group, the Chamberlain Pass Formation, is a thin (ca. 11 m) unit thought to have been deposited in the Middle to Late Eocene based on dates from the top of the overlying Chadron Formation of 33.9 ± 0.13 Ma (Prothero and Whittlesey, 1995). The formation comprises conglomerate, sandstone channel deposits, overbank mudstones, and floodplain paleosols (Terry and Evans, 1994).

Paleosols of the Chamberlain Pass Formation are reported to contain root traces inside and outside BADL (Retallack, 1983; Tweet et al., 2011), and large pieces of fossil wood tentatively identified as Cupressaceae have been found in a basal conglomerate of the formation east of the park (Clark, 1967).

Chadron Formation: The overlying Chadron Formation is interpreted to represent stream, pond, riverbank, and floodplain deposits that document a transition from a humid or subhumid woodland environment to drier and cooler grasslands during the latest Eocene (ca. 37–34 Ma) (Retallack, 1983; Prothero and Emry, 2004). Paleobotanically productive strata in this formation are the Gleska and Ohaka Series paleosols and a level at the base of the formation that produced fossil wood. Root traces occur in both the Gleska and Ohaka series paleosols, and the Gleska series is also a known source of *Celtis* endocarps (Retallack, 1983). The woody material collected from the base of the formation is a rooted stump with potential affinity to the Dipterocarpaceae, a tropical plant family with a modern center of diversity in southeast Asia (Shuler and Santucci, 1986). Outside of BADL, the Chadron Formation contains freshwater limestone deposits with charophytes and aquatic plants and vertebrate coprolites with unidentified plant remains (Evans, 1998, 1999; Sandler et al., 1998).

Brule Formation: The youngest member of the White River Group is the Lower Oligocene (ca. 33–30 Ma) Brule Formation (Prothero and Emry, 2004). The Brule Formation is divided into two members. The lower Scenic Member is characterized by well-developed and highly fossiliferous pinkish and green-gray paleosols and lesser quantities of silt, sand, and volcanic ash; the upper Poleslide Member is characterized by light-gray to gray-brown paleosols, eolian volcanic sediments, and highly fossiliferous sandstones (Tweet et al., 2011). The paleosols of each member contain root traces and *Celtis* endocarps and are interpreted as documenting a continued transition from a warm temperate and subhumid climate to a cool temperate and semiarid climate (Retallack, 1983, 1984). The Scenic Member of the Brule Formation also preserves unidentified tree stumps and bits of plants in coprolites (DiBenedetto, 1998; Wrenn and DiBenedetto, 2002).

Sharps Formation: The Upper Oligocene Sharps Formation overlies the White River Group and also comprises floodplain deposits and extensive paleosols (Tweet et al., 2011). The formation is divided by an unconformity into a lower (ca. 29.9–28.9 Ma) and an upper (ca. 28.4–20.0 Ma) section. This formation contains several series of paleosols that have been used to infer a cool semiarid climate and steppe/prairie vegetation (Retallack, 1981). The only paleobotanical material to be reported from the Sharps Formation within BADL are root traces (Retallack, 1983).

Quaternary sediments: Rawling and Fredlund (1999) mention undescribed phytoliths collected from soils buried under eolian dunes at Sheep Mountain Table in BADL. Quaternary floodplain, river terrace, eolian, landslide, and alluvial deposits are present elsewhere in the park and may contain similar remains.

Agate Fossil Beds National Monument

Agate Fossil Beds National Monument (AGFO), located in northwestern Nebraska, is known for its rich vertebrate remains within fluvial Miocene units (Kottkamp et al., 2020). These units also preserve evidence of ancient plant communities in root traces (rhizoliths) and plant fragments recovered from the Harrison and Anderson Ranch formations.

Harrison Formation: The Lower Miocene Harrison Formation is part of the Arikaree Group, a sedimentary package characterized by eolian, volcaniclastic, and fluvial sediments deposited during the Late Oligocene and Early Miocene (Hunt, 1990; Kottkamp et al., 2020). Paleosols of the Harrison Formation preserve evidence of ancient vegetation in root traces and mollic epipedons, surface horizons of small, rounded clods of clayey and organic-rich soil that are produced by the fine roots and abundant earthworms characteristic of grassland ecosystems (Retallack, 1997; Kottkamp et al., 2020). Some root traces are associated with plant fragments and occur within the corkscrew-shaped burrows of an extinct beaver called *Palaeocastor* (Martin and Bennett, 1977). These large burrows (210–275 cm [83–108 in] in length), known as *Daemonelix*, were initially interpreted as freshwater sponges or a novel type of extinct plant (Martin and Bennett, 1977). The age of the Harrison Formation is constrained to 22.9 ± 0.08 Ma based on $^{40}Ar/^{39}Ar$ dates of the Agate Ash, a thinly laminated white tuff that has been diagenetically altered to smectite (Izett and Obradovich, 2001; Kottkamp et al., 2020).

Anderson Ranch Formation: The Anderson Ranch Formation is fossiliferous throughout AGFO and contains two particularly well-known bonebeds at Carnegie Hill and University Hill. Each of these localities contains a succession of fluvial, lacustrine, eolian/volcaniclastic, and shallow freshwater deposits (Kottkamp et al., 2020). Rhizoliths are present in silty sandstone deposits containing eolian volcaniclastic loess (Hunt, 1990). The conformable contact of the Anderson Ranch Formation with the underlying Harrison Formation is relatively close to the Agate Ash, suggesting that sediments of the Anderson Ranch Formation were also deposited during the Early Miocene, not much later than 22 Ma (Kottkamp et al., 2020).

Scotts Bluff National Monument

Scotts Bluff National Monument (SCBL) in western Nebraska was established to preserve Scotts Bluff, a landmark for the California, Oregon, and Mormon trails, and the surrounding ca. 3,000 acres (Tweet et al., 2011). Paleobotanical material found within the monument is mainly confined to root traces within paleosols, but equivalent units to those inside the park contain root traces, plant fragments, phytoliths, *Celtis* endocarps, and pollen.

Brule Formation: The Lower Oligocene Brule Formation is the uppermost unit of the White River Group. Exposures of the Brule Formation within SCBL are divided into two members, the Orella Member and the Whitney Member, and these members are correlative with the Scenic and Poleslide members of the Brule Formation recognized in BADL (Tweet et al., 2011). Each member is known to contain paleobotanical resources near SCBL, but material collected from these units within SCBL is poorly documented. The Orella Member is the basal unit of the Brule Formation at SCBL and is composed of brown, gray, green, and pink volcaniclastic mudstone and siltstone (Tweet et al., 2011). Outside of the park, the Orella Member has produced phytoliths (Strömberg, 2004), root traces (LaGarry et al., 1998), casts of tree limbs, and other plant fragments (Nixon and LaGarry-Guyon, 1993). The age of these resources is constrained to 34–31.85 Ma based on correlations of the underlying Chadron Formation and an ash layer within the overlying Whitney Member (Tweet et al., 2011). The Whitney Member has also produced paleobotanical material outside of SCBL, including plant fragments, root traces, phytoliths (Meehan, 1994; Grandstaff and Terry, 2009), and *Celtis* endocarps (Ash Hollow State Park; Diffendal et al., 1996). Based on the age of the base of the unconformably overlying Gering Formation, these materials were deposited between 31.85 and 28.31 Ma (Swinehart and Diffendal, 1997).

Arikaree Group: Within SCBL, the Arikaree Group is represented by the Upper Oligocene Gering Formation, and the undivided Upper Oligocene to Lower Miocene Monroe Creek and Harrison formations. The Gering Formation is correlative with the Upper Sharps Formation exposed in BADL, and it also preserves root traces (Loope et al., 2005). Deposition of the Gering Formation is thought to have occurred in a relatively short period of time based on an age of 28.1 ± 0.03 Ma for the Twin Sisters Pumice Conglomerate low in the unit, and an age of 27.29 ± 0.08 Ma for an ash bed in the overlying Monroe Creek Formation (Tedford et al., 1996; Bailey, 2004). The undivided rocks of the Monroe Creek Formation and Harrison Formation are brown and gray silty sandstones representing deposition in braided streams, ephemeral shallow lakes, and interchannel plains blanketed by windblown sediments. Within

SCBL, these rocks contain zones of root traces and bioturbation (Graham, 2009; Tweet et al., 2011), and phytoliths and root traces have been found in these formations outside of the park (Strömberg, 2002, 2004).

Chimney Rock National Monument

The Chimney Rock National Historical (CHRO) Site in southwestern Nebraska preserves a prominent landmark (Chimney Rock) for the California, Mormon, and Oregon trails (Tweet et al., 2011). The site, an NPS-affiliate owned by the Nebraska State Historical Society, is not officially part of the NGPN but has similar geology to SCBL located just 30 km (19 mi) to the northwest. The only paleobotanical materials preserved at CHRO are root casts in the Whitney Member of the Oligocene Brule Formation (Meehan, 1994). Outside of CHRO, the Whitney Member preserves plant fragments, root traces, phytoliths (Meehan, 1994; Grandstaff and Terry, 2009), and *Celtis* endocarps (Ash Hollow State Park; Diffendal et al., 1996).

Niobrara National Scenic River

Niobrara National Scenic River (NIOB) protects an area of about 23,000 acres along a 122 km (76 mi) stretch of the Niobrara River in north-central Nebraska. This NPS unit is rich in paleontological resources and preserves an extensive sequence from the past 15 million years (Tweet et al., 2011). The Ogallala Group, comprised by the Valentine and Ash Hollow formations in NIOB, represents the Miocene portion of this sequence and preserves a variety of plant fossils.

Valentine Formation: The Valentine Formation underlies the Ash Hollow Formation, and its age is constrained to 13.6–12.18 Ma by the Hurlburt Ash of the basal Cornell Dam Member of the Valentine Formation and the Swallow Ash of the basal Cap Rock Member of the Ash Hollow Formation (Woodburne, 2007). Within NIOB, the Cornell Dam Member has produced fossil reproductive structures of *Berriochloa* and *Celtis* alongside abundant mammal and herpetological remains at the Norden Bridge and Egelhoff quarries (Voorhies and Corner, 1993; Gabel et al., 1998); wood is also present at Norden Bridge and other NIOB sites in the Valentine Formation (Voorhies, 1990). Fossil wood (Figure 17) has been reported from the overlying Crookston Bridge Member of the Valentine Formation (G. Warrick, NIOB, written comm. to Tweet and Santucci, August 2021). One well-preserved piece of wood has vessels and aliform parenchyma; This combination of traits suggests an affinity with either the legume (Fabaceae) or mulberry (Moraceae) family. Other logs and wood fragments have been observed eroding from sandy channel deposits around the level of the Niobrara River. Outside of NIOB, the Kilgore flora contains a rich assemblage of similarly aged leaf and fruit fossils dominated by the leaves and associated fruitlets of *Platanus vitifolia* (MacGinitie, 1962).

Ash Hollow Formation: The Ash Hollow Formation documents deposition spanning at least 12.18–6.62 Ma based on ashes dated from the lower and upper parts of the unit (Tucker and Voorhies, 2005; Tweet et al., 2011). The only paleobotanical material recognized from the Ash Hollow Formation within NIOB are root casts in the Cap Rock Member, but the formation contains diverse plant records outside of the park. Three-dimensional, occasionally silicified fruits and seeds of Poaceae, Cyperaceae, Juglandaceae, and Ulmaceae have been recovered from several horizons of the Ash Hollow Formation during field collections and by floating sediments in water (Thomasson, 1980, 1987; Gabel et al., 1998). One particularly notable finding from the Cap Rock Member outside of NIOB is the preservation of *Berriochloa* anthoecia within the oral cavity and rib cage of articulated *Teleoceras* specimens (Voorhies and Thomasson, 1979). This record provides insight into the diet of the extinct mammal and trophic interactions in the Miocene grassland of the Great Plains.

3.4.2 Great Lakes Inventory and Monitoring Network

Ice Age National Scenic Trail and Ice Age Scientific Reserve

Ice Age National Scenic Trails traces the southern boundary of the last glacial maximum for 2,000 km (about 1,200 mi) between Taylors Falls and Sturgeon Bay, Wisconsin. Ice Age National Scientific Reserve is an NPS-affiliate comprising nine Ice Age related areas managed by the state of Wisconsin. Paleobotanical resources within Ice Age National Scenic Trail areas include Late Pleistocene and Holocene palynomorphs from Devil's Lake State Park (Attig et al., 1990), Butler Lake (Long et al., 2011), and the Two Creek Forest Bed in Two Creeks Buried Forest State Natural Area. The Two Creek Forest Bed is a 0.5–5 cm (0.2–2 in) thick layer of organic debris and macroscopic plant parts dated to 11,850 ybp (Culberson, 1955; Broecker and Farrand, 1963). The plant and insect communities of the Two Creek Forest Bed include 19 species of bryophytes, among them two (*Ditrichum flexicaule* and *Calliergoa turgescens*) that no longer occur in Wisconsin; wood, twigs, and cones of *Picea mariana;* and coleopteran, dipteran, and hymenopteran insects (Wilson, 1932, 1936; Culberson, 1955; Morgan and Morgan, 1979).

Pictured Rocks National Lakeshore

Pictured Rocks National Lakeshore (PIRO) protects 730,000 acres on the southern shore of Lake Superior in Michigan's Upper Peninsula. Proterozoic and Paleozoic sandstones comprise much of the bedrock within the lakeshore, but Holocene paleobotanical material is preserved in glaciofluvial deposits and ancient soils buried by the Great Sable Sand Dunes (Anderton and Loope, 1995), and in sediment cores drawn from Great Sable Lake and Beaver Lake (Fisher and Whitman, 1999; Loope et al., 2004). The macropaleobotanical materials recovered from these localities are wood fragments that have been radiocarbon dated to establish a chronology for organic horizons within sediment cores (Anderton and Loope, 1995; Fisher and Whitman, 1999; Loope et al., 2004). In sediment cores from Great Sable Lake, Loope et al. (2004) identified a buried northern white cedar (*Thuja occidentalis*) forest from ca. 2990 ybp. Outside of PIRO, fossil pollen from the early Holocene (9500–6000 ybp) has been recovered from Buck Lake sediments and includes *Picea*, *Pinus*, *Betula*, *Acer*, *Quercus*, *Ulmus*, other deciduous trees, and prairie forbs (Davis et al., 2000).

3.4.3 Heartland Inventory and Monitoring Network

Herbert Hoover National Historic Site

Herbert Hoover National Historic Site (HEHO) in eastern Iowa encompasses about 200 acres of surficial, unconsolidated Quaternary sediments (Graham, 2011). Known Cenozoic paleobotanical resources in HEHO are limited; the only plant fossil reported from HEHO is a piece of Pleistocene petrified wood that was found 8 m (26 ft) below the surface in a well dug in 1983 (Graham, 2011, 2017).

3.5 Southeast, National Capital, and Northeast Regions

The Southeast, National Capital, and Northeast regions encompass the remainder of NPS units that contain Cenozoic paleobotanical resources (Figure 18). Cenozoic paleobotanical material is comparatively sparse in this part of the USA but has been found in 15 NPS-administered units that fall into six inventory and monitoring networks: the Gulf Coast Network, National Capital Region Network, Mid-Atlantic Network, Eastern Rivers and Mountains Network, Northern Coast and Barrier Network, and the Northeast Temperate Network.

Paleobotanical resources from the NPS units in eastern USA mainly occur in surficial Quaternary deposits or in the subsurface; however, numerous significant paleofloras are known from outside NPS areas. Paleogene floras of the

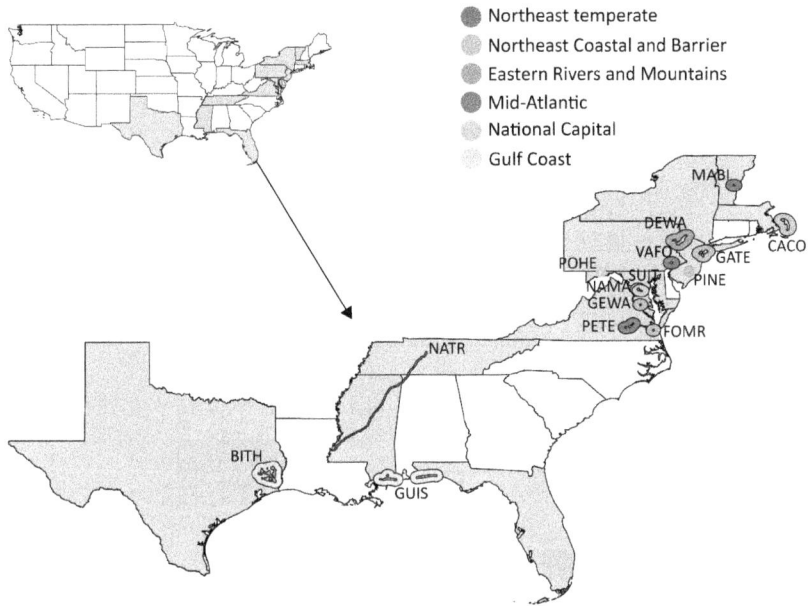

Figure 18 Map of NPS units with Cenozoic paleobotanical resources in the Southeast, National Capital, and Northeast regions, colored by inventory and monitoring network. See Figure 1 legend for details on the park symbology and acronyms.

region are preserved in sediments of the Wilcox and Claiborne Groups in the Mississippian Embayment (Berry, 1916a; Dilcher, 1971; Crepet and Nixon, 1989), and a smaller number of Neogene floras are scattered across the eastern USA in Tennessee (Gray fossil site; Hermsen, 2021, 2023), Alabama (Citronelle flora; Stults and Axsmith, 2015), Maryland (Brandywine flora; McCartan et al., 1990), and Vermont (Brandon Lignite; Tiffney, 1994). Within NPS units, some notable Quaternary assemblages include the Pleistocene seed and fruit floras found alongside abundant vertebrate remains in the Cumberland and Port Kennedy bone caves, which are located along the Potomac National Heritage Trail and in Valley Forge National Historic Park, and the remains of a Pleistocene bald cypress (*Taxodium*) swamp discovered during excavations at Washington Circle, in the National Mall, Washington, DC (Figure 19). Notable subsurface discoveries from NPS units are the palynofloras of the Mays Landing borehole from affiliated Pinelands National Preserve in New Jersey that provide data on the composition of Oligocene and Miocene floras of the northeast (Owens et al., 1988; Graham, 1999) and the palynofloras of Duck Pond from Cape Cod National Seashore that record changing tree distributions during the Holocene (Winkler, 1985).

Park Unit	Significance	PA	E	O	M	PL	Q
MABI							■
CACO				■			■
GATE							■
PINE	■			■			■
GEWA					■		■
FOMR							■
DEWA							■
VAFO							■
PETE							■
POHE							■
NAMA							■
SUIT						■	
NATR		■					
BITH					■	■	■
GUIS						■	

■ Confirmed   ■ Unconfirmed
■ Highest ■ High ■ Moderate ■ Present

Figure 19 Summary of the temporal distribution and significance of Cenozoic paleobotanical resources in the Southeast, National Capital, and Northeast regions. See Figure 2 legend for explanation of significance ratings.

3.5.1 Northeast Temperate Inventory and Monitoring Network

Marsh-Billings-Rockefeller National Historic Park

Marsh-Billings-Rockefeller National Historic Park (MABI) in eastern Vermont commemorates an area significant to the lives of several conservationists: George Perkins Marsh, Frederick Billings, Mary French Rockefeller, and Laurance S. Rockefeller. The only Cenozoic paleobotanical materials known from within MABI are two pieces of marl from The Pogue, a former wetland that is now a manmade pond (Tweet et al., 2010; Thornberry-Ehrlich, 2011). These pieces of marl are dated to the Holocene and preserve veined leaf impressions, twigs, and grass stems (Tweet et al., 2010; Thornberry-Ehrlich, 2011). The specimens are held in the park's collections.

3.5.2 Northeast Coastal and Barrier Inventory and Monitoring Network

Cape Cod National Seashore

Cape Cod National Seashore (CACO) in Massachusetts preserves about 65 km (40 mi) of sandy beaches, marshes, and ponds along the Atlantic coast. Paleobotanical material has been recovered from CACO in four different contexts: microfossils in subsurface deposits retrieved from deep wells or boreholes; reworked petrified wood in surficial outwash deposits; Pleistocene petrified wood; and microfossils in surficial lake or pond sediments retrieved by shallow cores (Tweet et al., 2014). The material found in deep wells from CACO are palynomorphs, lignite coal, and carbonized wood interpreted as Eocene in age (Zeigler et al., 1964), and the fossiliferous sediments may have been initially deposited elsewhere and subsequently transported by glaciers (Uchupi et al., 1996). Dark gray silicified wood fragments that are Cretaceous or Early Cenozoic in age are present in the glacial drift of western Cape Cod and have traced to a bedrock source near Middleboro, Massachusetts (Oldale, 2001). Several samples of Pleistocene petrified wood were found in the Wellfleet Quadrangle, within CACO and dated to over 25,000 radiocarbon ybp (Oldale, 1968). Quaternary pollen cores from lakes and ponds within CACO are the park's most significant paleobotanical material and record the regional vegetation over approximately the past 11,500 years (Tweet et al., 2014). One site in Wellfleet, Duck Pond, has an especially complete record of the past 12,000 years, and its deposits include identifiable moss macrofossils (Winkler, 1985).

Gateway National Recreation Area

The Gateway National Recreation Area (GATE) consists of three geographically isolated units in New York (NY) and New Jersey (NJ): Jamaica Bay Unit (NY), Sandy Hook Unit (NJ), and the Staten Island Unit (NY). Pre-Quaternary deposits are not exposed in any of these units, and there is a hiatus between the Upper Cretaceous and Pleistocene in the bedrock (Tweet et al., 2014), limiting the potential of GATE to furnish Cenozoic plant macrofossils.

Paleobotanical resources discovered within or near GATE are specimens of plant debris found in reworked nodules that have washed ashore on beaches of the Sandy Hook and Jamaica Bay units (Chamberlain et al., 2006) and lignite, wood, rootlets, and reed fragments recovered from boreholes at the Sandy Hook and Jamaica Bay units (Tweet et al., 2014). Potential sources of fossiliferous nodules are the Quaternary Cape May (300–75 kya) and Gardiner Clay (125–75 kya) formations and the Cenomanian–Turonian Magothy and Raritan formations (Chamberlain et al., 2006; Tweet et al., 2014).

Boreholes from the Sandy Hook Unit extend through Quaternary and Upper Cretaceous deposits (Kümmel and Poland, 1910; Minard, 1969). Some of the Cretaceous deposits have been attributed to the Raritan Formation, which has yielded exceptional plant macrofossils and mesofossils outside of GATE (Berry, 1911; Crepet et al., 2001). Higher in the stratigraphic column, Quaternary pollen records from GATE are relevant to reconstructions of forest succession in New York and New England during the Holocene (Tweet et al., 2014).

Pinelands National Reserve

Pinelands National Reserve (PINE) is an NPS-affiliated unit located in the New Jersey Coastal Plain. As with other units of the eastern states, most paleobotanical resources and all pre-Quaternary resources associated with the Pinelands have been derived from boreholes. Boreholes taken from several localities within PINE (Owens et al., 1988; Miller et al., 1994, 1998, 1999; Browning et al., 2011) provide a detailed account of the region's subsurface stratigraphy and include paleobotanical materials such as pollen, wood, and phytoliths from formations that span the Middle Eocene to Late Miocene. The palynofloras of the Mays Landing borehole (Owens et al., 1988) are one of few available lines of evidence for reconstructing the Paleogene and Neogene vegetational history of the northeastern USA (Graham, 1999). The Pinelands are also a source of Quaternary palynological resources, and pollen samples have been used to study the dynamics of forest succession during the Pleistocene and Holocene (Potzger, 1945; Florer, 1972).

Paleogene units: The oldest Cenozoic paleobotanical material discovered within PINE are wood fragments and burrowed pyritized wood from the Middle–Late Eocene Shark River Formation; these were retrieved from corehole ACGS-4 along Great Egg Harbor River and the Double Trouble core site located in the northern part of PINE (Owens et al., 1988; Browning et al., 2011). The Shark River Formation represents lower shoreface and marine shelf deposits. Borehole ACGS-4 also produced wood and pollen from three Oligocene units representing nearshore and marine shelf deposits (Owens et al., 1988).

Cohansey Sand and Kirkwood Formation (Miocene): The Cohansey Sand was deposited during the Middle to Late Miocene in paludal and lagoon environments, and an organic-rich layer present in the lagoon deposits has yielded wood fragments and pollen in boreholes taken along Great Egg Harbor River (ACGS-4 corehole) and Bass River (Owens et al., 1988; Miller et al., 1998). The Kirkwood Formation represents neritic and prodeltaic deposits from the Early to Middle Miocene, and sections from corehole ACGS-4 and a core taken in Ancora, New Jersey, contain wood, pollen, and phytoliths (Owens et al., 1988; Miller et al., 1999).

George Washington Birthplace National Monument

George Washington Birthplace National Monument (GEWA) is located in eastern Virginia along the Potomac River and preserves the area around the former plantation owned by the Washington family. Plant macrofossils have been recovered in GEWA from the Lower to Middle Miocene Calvert

Formation and just outside of GEWA from the Pleistocene Tabb Formation. These units are exposed, respectively, on riverbanks of the Potomac and one of its tributaries, Popes Creek.

Calvert Formation: The Lower to Middle Miocene Calvert Formation represents shallow marine deposits and is known to contain diverse macrofossil types such as wood, leaves, invertebrate shell fragments, fish bones, scales, teeth, and mammal bones (Shattuck, 1904; Vogt and Eshelman, 1987; Tweet et al., 2014; Tweet and Santucci, 2017). The exposures within GEWA are a source of shark teeth and pieces of carbonized wood, presumably derived from waterlogged trunks swept out to sea. Significant plant macrofossil localities of the Calvert Formation occur in Richmond, Virginia, and the District of Columbia, and are interpreted as representing the landward extent of the formation (Berry, 1916b).

Tabb Formation: The Pleistocene Tabb Formation consists of river and coastal plain deposits and is subdivided into the Sedgefield, Lynnfield, and Poquoson members, which collectively span from 130 to 33 kya (Mixon et al., 1989; Scott et al., 2010). An assemblage of fossil fruits, seeds, and tendrils was recovered from deposits of the Tabb Formation on the bank of Popes Creek outside of GEWA, a tributary of the Potomac River that forms the park's southern boundary (Berry, 1938; Tweet et al., 2014). Taxa described from this assemblage include aquatic monocotyledons (*Naias* and *Zannichellia palustris*) and herbaceous, woody, and vining dicotyledons (*Quercus*, *Phytolacca*, *Vitis*) (Berry, 1938). Plant fossil localities of the Tabb Formation have also been recovered to the west and south of GEWA on the banks of the Rappahannock River (Berry, 1906, 1909). At these localities, plant macrofossils – mainly seeds – were recovered from a peat layer overlying *in situ* cypress stumps and likely come from the Sedgefield Member of the Tabb Formation (Mixon et al., 1989).

Fort Monroe National Monument

Fort Monroe National Monument (FOMR) commemorates a historic military base in Hampton, Virginia, on the southern tip of the Virginia peninsula. From an artesian well in FOMR, Fontaine (1882) described a 1 m (3 ft) stratum of coarse, yellowish-green sand and indurated sandstone containing coal and plant fragments intervening between Miocene marine deposits above and below.

3.5.3 Eastern Rivers and Mountains Inventory and Monitoring Network

Delaware Water Gap National Recreation Area

Delaware Water Gap National Recreation Area (DEWA) preserves about 70,000 acres of forest and waterway along the Delaware River in northwestern

New Jersey and southwestern Pennsylvania. The only paleobotanical discovery within Cenozoic sediments of DEWA is sub-fossil wood recovered at a depth of 6.4 m (21 ft) in a pit dug during the construction of the Bushkill access area in Bushkill, Pennsylvania. The log yielded a radiocarbon date of 4105 ± 90 ybp (ver Straeten et al., 2001).

3.5.4 Mid-Atlantic Inventory and Monitoring Network

Valley Forge National Historic Park

Valley Forge National Historic Park in eastern Pennsylvania preserves 3,500 acres surrounding an encampment site of the Continental Army during the Revolutionary War. The area surrounding the former encampment includes the Kennedy Bone Cave, a sinkhole deposit that was discovered in the 1870s during limestone quarrying near the former town of Port Kennedy, Pennsylvania. The cave is thought to have formed as an underground cavern from dissolution of limestone, and it eventually became a sinkhole when the roof collapsed. During the middle Pleistocene, amphibians, reptiles, birds, mammals, and plants fell into the hole and became buried by debris. Plant remains include twigs, leaves, fruits, and seeds and are referrable to extant species, including *Quercus alba*, *Corylus americana*, and *Pinus rigida* (Mercer, 1899).

Petersburg National Battlefield

Petersburg National Battlefield in Petersburg, Virginia, commemorates the site of a prolonged siege between the Confederate and Union armies during the American Civil War. The unit comprises a small area with few geologic exposures, but a twig was reported in an oxbow lake clay deposit from the center of the Petersburg Crater site (Hal Pranger, Geologic Resources Division, pers. comm.).

3.5.5 National Capital Region Inventory and Monitoring Network

Potomac Heritage National Scenic Trail

Potomac Heritage Trail (POHE) encompasses about 1,100 km (684 miles) of existing and planned trails that traverse Pennsylvania, Maryland, Virginia, and Washington, DC. The Cumberland Bone Cave in Cumberland, Maryland, is a sinkhole deposit that preserves a middle Pleistocene flora and fauna and is located along the Great Allegheny Passage trail of the POHE network (Hodnett et al., 2023a). The cave was collected periodically between 1912 and 2012 and contains snails, fish, amphibians, reptiles, birds, mammals, and plants. Plant remains include indeterminate seeds and grass stems and fruits of *Celtis occidentalis* and *Quercus*.

National Mall and Memorial Parks

The National Mall and Memorial Parks (NAMA) in Washington, DC, includes fourteen NPS units that preserve 1,000 acres of parkland and numerous significant natural and cultural resources within the nation's capital (Figure 18). Pleistocene fossil plant material has been recovered from two units of NAMA, the Washington Monument and Washington Circle (Darton, 1950; Knox, 1966). The fossil plant material from the Washington Monument was recovered from three of fifteen test boreholes that were made to evaluate the prospect of placing concrete piers down to bedrock under the foundation of the monument. These cores revealed that approximately 18 m (60 ft) of sand, gravel, and blue clay overly bedrock within the grounds of the Washington Monument and yielded partially petrified wood and petrified wood in sand (Darton, 1950).

Various excavations in the District of Columbia between 16th Street NW and Washington Circle have revealed deposits of a Pleistocene bald cypress swamp (Berry, 1933; Knox, 1966). These deposits contain *in situ* cypress stumps and diverse fruits, seeds, leaves, and pollen (Berry, 1933; Knox, 1966). Berry (1933) described 37 different kinds of plants, which include bald cypress (*Taxodium distichum*), monocots, and woody and herbaceous dicots. Knox (1966) mapped the lateral extent of the swamp and collected pollen samples from several levels of the swamp deposits. The dominant components of the palynoflora shift from spruce to pine to deciduous angiosperms moving up in the profile, and Knox (1966) suggested that these shifts reflect a warming trend that is concordant with the swamp having existed during the most recent (Sangamon) interglacial, approximately 100,000 ybp.

Suitland National Parkway

Suitland National Parkway (SUIT) is part of National Capital Parks East (NACE), a management unit that encompasses 14 park sites, parkways, and statuary preserving significant natural and cultural heritage and covering about 8,000 acres between Capitol Hill and the Maryland suburbs (Nelson et al., 2019). Underlying this network are three significant fossiliferous rock units: the Potomac Group (Upper Cretaceous), the Pamunkey Group (Paleocene–Eocene), and the Chesapeake Group (Lower–Middle Miocene). Cenozoic plant macrofossil occurrences in this network are limited to undiagnostic wood fragments recovered from within SUIT during the construction of the Branch Avenue Metro Station (Kenworthy and Santucci, 2004). However, plant fossil localities in close proximity to NACE units have been recorded in the Calvert Formation at Good Hope Hill, Washington, DC, and in "Upland gravels" (map unit Tu) in Brandywine, Maryland (Berry, 1916b; Carr, 1950; McCartan et al., 1990). The Calvert Formation consists mainly of shallow marine sediments deposited during the Early and Middle Miocene (Vogt and Eshelman, 1987; Tweet et al., 2014). It is present within several NACE units and is a major formation within SUIT, where it has yielded cetacean and other mammal fossils (Nelson et al., 2019).

The "Upland gravels" are an informal unit consisting of poorly sorted sand, cobbles, pebbles, and gravel laid down by the ancestral Potomac River during the Late Miocene and Pliocene (McCartan et al., 1990; Nelson et al., 2019), and the deposits are present with six NACE units. This unit is the source of the Brandywine flora, one of few Neogene sites in eastern North America that has produced well-preserved macrofossils, including leaves, seeds, fruit, and pollen that represent 49 different taxa, and one of the last occurrence of *Pterocarya* (Juglandaceae) before it went extinct in North America (McCartan et al., 1990; Stults et al., 2011, 2022).

3.5.6 Gulf Coast Inventory and Monitoring Network

Natchez Trace Parkway

Natchez Trace Parkway (NATR) is a 715 km (444 mi) recreational road that stretches between Natchez, Mississippi, and Nashville, Tennessee. It does not contain confirmed records of Cenozoic paleobotanical material; however, leaf impressions and fossil wood have been found from a locality known as the Jeff Busby site in the Little Mountain Recreation Area, near Mathiston, Mississippi (Black, 1962; Kenworthy et al., 2007). Based on geological maps, this material was likely collected from the Paleocene Tuscahoma Formation, but the locality has not been relocated since it was first found and it is not clear if the fossils came from within or just outside NATR (Hodnett et al., 2023b). There are other Cenozoic formations present within NATR that have yielded paleobotanical fossils outside the park boundary. These formations include the Zilpha Clay, Porters Creek Clay, Kosciusko Sand, and Vicksburg Group (Kenworthy et al., 2007).

Outside of NATR, the Tuscahoma formation has produced Paleocene and Eocene palynomorphs from cores in Mississippi and Alabama (Harrington, 2008) and an Early Eocene macroflora from the Red Hot Truck Stop locality in Meridian, Mississippi (Danehy et al., 2007). This formation is thought to include the Paleocene–Eocene boundary (Harrington et al., 2004).

Big Thicket National Preserve

Big Thicket National Preserve (BITH) in east Texas was the first preserve to be incorporated in the NPS, in 1974. The preserve protects water corridors and a variety of plant communities, such as floodplain forest and swamps, savannah, and upland forests, that comprise approximately 97,000 acres. Within BITH, geologic strata are mainly exposed in erosive contexts such as riverbanks. They were deposited between the Miocene and the Holocene (Kenworthy et al., 2007). Three formations within BITH, the Miocene Fleming Formation, the Pliocene–Pleistocene Willis Formation, and the Pleistocene Beaumont Formation, have produced petrified wood, but in some instances, this wood may be reworked from older sources exposed north of the park, such as the Catahoula Formation (Singleton, 2008; Thornberry-Ehrlich, 2018).

The Willis Formation is the most significant source of petrified wood within BITH. Fossil localities from this formation along the Trinity and Neches rivers yield petrified wood from basal gravels and from Pleistocene fluvial strata (Kenworthy et al., 2007). The wood specimens from localities in the park are unstudied, but areas adjacent to BITH with Pleistocene outcrops of the Willis Formation have produced fossil wood belonging to *Quercus* and *Ulmus* (Singleton, 2008; Thornberry-Ehrlich, 2018). In the northern section of BITH, a piece of petrified palm wood was reported, but the record lacks stratigraphic context (Watson, 2006).

Gulf Islands National Seashore

Gulf Islands National Seashore (GUIS) protects 258 km (160 mi) of coastline, primarily barrier islands off the mainland of Florida and Mississippi. The only paleobotanical record from GUIS is pollen and pulverized plant material recovered from Pliocene sediments in a core taken from Horn Island in Mississippi (Gohn et al., 1996; Clinton et al., 2023).

4 Conclusion

In our inventory report, we have demonstrated that the fossil plant resources contained in the NPS are: (1) geographically widespread, occurring in 74 NPS units distributed through all seven NPS regions and 25 states; (2) temporally widespread, occurring in 13 Paleocene formations, 27 Eocene formations, 19 Oligocene formations, 27 Miocene formations, 12 Pliocene formations, and 50 Quaternary units; (3) taxonomically diverse, including bryophytes, ferns, gymnosperms, and at least 40 families of angiosperms; (4) globally significant, yielding 393 type specimens and numerous biostratigraphically and biogeographically notable occurrences; and (5) in need of further curation, monitoring, and research.

Cenozoic fossil plants of the NPS are a significant scientific, educational, and cultural resource that have great potential to deepen global understanding of the history of plants, terrestrial ecosystems, and climate, and to connect and educate public audiences about their natural heritage. Cenozoic NPS paleobotanical resources are also heterogeneous, and vary considerably in their quality and abundance, extent of previous collection and study, and degree of awareness among scientific and public audiences. On one end of this spectrum are paleobotanical Lagerstätten such as the lacustrine shales of Florissant Fossil Beds and Fossil Butte National Monuments, and the fossil forests of Yellowstone National Park that have sustained over a century of paleobiological research and hold prominent profiles in the public's perception of paleobotany; even among these sites, a consummate baseline understanding of taxonomy has not been achieved (e.g., Supplementary Table 8). Then there are NPS units like Aniakchak National Monument and Preserve, Denali, Katmai, and Wrangell–St. Elias National Parks and Preserves that have produced extensive paleobotanical collections, including type specimens, but require additional geochronologic, sedimentological, and stratigraphic context as well as further taxonomic study to be fully integrated into our understanding of the floristic, vegetational, and paleoclimatic history of North America. Conversely, there are other NPS units such as Theodore Roosevelt and Glacier National Parks whose paleobotanical resources are well-situated in stratigraphic context but have not yet, or only recently (e.g., Smith et al., 2023) been extensively collected or studied taxonomically. Elsewhere on this spectrum are the late Neogene and Pleistocene resources of Golden Gate and Santa Monica National Recreation Areas that represent small, well-studied collections that are poorly known to the public despite being well-suited to science communication efforts. The records of these two parks document the recent evolutionary history of extant species (e.g., *P. radiata*) and ecosystems (e.g., closed-cone pine forests) with possible implications for land management and conservation (Axelrod, 1967; Millar, 1998). Finally, there are the many instances where NPS units contain minor paleobotanical resources (e.g., occasional petrified wood) that should continue to be monitored but are unlikely to provide major contributions to paleobotany.

The successful management of NPS fossil plant resources to fulfill their potential for scientific research and science communication will require an appropriate pairing between management intensity and the scope, significance, and distribution of fossil plants within NPS units. The baseline inventory presented here offers a starting point for the development of monitoring, curation, and research strategies that meet the unique needs of each of these 74 NPS units with Cenozoic paleobotanical resources. After fostering over 150 years of paleobotanical exploration, lands protected by the National Park Service continue to yield important fossil discoveries and will certainly play an important role in the future of North American paleobotany.

References

Addicott, W. O., Winkler, G. R., & Plafker, G. (1978). Preliminary megafossil biostratigraphy and correlation of stratigraphic sections in the Gulf of Alaska Tertiary province. *U.S. Geologic Survey Open File Report*, 78–491, 2 sheets. https://doi.org/10.3133/ofr78491.

Alvarez, L. W., Alvarez, W., Asaro, F., & Michel, H. V. (1980). Extraterrestrial cause for the Cretaceous-Tertiary extinction. *Science*, **208**(4448), 1095–1108. https://doi.org/10.1126/science.208.4448.1095.

Anderson, L. C., Hartman, J. H., & Wesselingh, F. P. (2006). Close evolutionary affinities between fresh-water corbulid bivalves from the Neogene of western Amazonia and Paleogene of the Northern Great Plains, USA. *Journal of South American Earth Sciences*, **21**(1–2), 28–48. https://doi.org/10.1016/j.jsames.2005.07.008.

Anderson, R. S., Hasbargen, J., Koehler, P. A., & Feiler, E. J. (1999). Late Wisconsin and Holocene subalpine forests of the Markagunt Plateau of Utah, southwestern Colorado Plateau, U.S.A. *Arctic, Antarctic, and Alpine Research*, **31**(4), 366–378. https://doi.org/10.1080/15230430.1999.12003321.

Anderton, J. B., & Loope, W. L. (1995). Buried soils in a perched dunefield as indicators of Late Holocene lake-level change in the Lake Superior Basin. *Quaternary Research*, **44**, 190–199. https://doi.org/10.1006/qres.1995.1063.

Andrews, H. N. (1939). Notes on the fossil flora of Yellowstone National Park with particular reference to the Gallatin Region. *American Midland Naturalist*, **21**(2), 454–460. https://doi.org/10.2307/2420547.

Andrews, H. N., & Lenz, L. W. (1946). The Gallatin Fossil Forest. *Annals of the Missouri Botanical Garden*, **33**(3), 309–313. https://doi.org/10.2307/2394453.

Ashlee, T. R. (1932). A contribution to the Latah Flora of Idaho. *Northwest Science*, **6**(2), 69–82.

Attig, J. W, Clayton, L., Lange, K. I., & Maher, L. J. (1990). The Ice Age geology of Devils Lake State Park. *Wisconsin Geological and Natural History Survey Educational Series*, **35**, 1–28.

Axelrod, D. I. (1940). A record of *Lyonothamnus* in Death Valley, California. *Journal of Geology*, **48**(5), 526–531. https://doi.org/10.1086/624907.

Axelrod, D. I. (1967). Evolution of the Californian closed-cone pine forest. In Philbrick, R. N., ed., *Proceedings of the Symposium on the Biology of the California Islands*. Santa Barbara, CA: Santa Barbara Botanic Garden, pp. 93–149.

Axelrod, D. I. (1971). Fossil plants from the San Francisco Bay Region. In Lipps, J. H., & Moores, E. M., eds., *Geologic Guide to the Northern Coast Ranges, Point Reyes Region, California*. Sacramento, CA: Geological Society of Sacramento, pp. 74–86.

Axelrod, D. I. (1986). Cenozoic history of some western American pines. *Annals of the Missouri Botanical Garden*, **73**(3), 565–641.

Axelrod, D. I. (1990). Ecologic differences have separated *Pinus remorata* and *P. muricata* since the Early Pleistocene. *American Journal of Botany*, **77**(3), 289–294. https://doi.org/10.1002/j.1537-2197.1990.tb13557.x.

Bacon, C. R. (2008). Geologic map of Mount Mazama and Crater Lake Caldera, Oregon. *U.S. Geological Survey Scientific Investigation Map*, **2832**, 1–45, 4 sheets, scale 1:24,000. https://doi.org/10.3133/sim2832.

Bailey, B. E. (2004). Biostratigraphy and biochronology of early Arikareean through late Hemingfordian small mammal faunas from the Nebraska Panhandle and adjacent areas. *Paludicola*, **4**(3), 81–113.

Baker, R. G. (1970). Pollen sequence from late Quaternary sediments in Yellowstone Park. *Science*, **168**(3938), 1449–1450. https://doi.org/10.1126/science.168.3938.1449.

Baker, R. G. (1976). Late Quaternary vegetation history of the Yellowstone Lake Basin, Wyoming. *U.S. Geological Survey Professional Paper*, **729-E**, E1–E48. https://doi.org/10.3133/pp729E.

Baker, R. G. (1986). Sangamonian(?) and Wisconsinan paleoenvironments in Yellowstone National Park. *Geological Society of America Bulletin* **97**(6), 717–736. https://doi.org/10.1130/0016-7606(1986)97%3C717:SAWPIY%3E2.0.CO;2.

Baker, R. G., & Richmond, G. M. (1978). Geology, palynology, and climatic significance of two pre-Pinedale lake sediment sequences in and near Yellowstone National Park. *Quaternary Research* **10**(2), 226–240. https://doi.org/10.1016/0033-5894(78)90103-5.

Barclay, R. S., Johnson, K. R., Betterton, W. J., & Dilcher, D. L. (2003). Stratigraphy and megaflora of a K–T boundary section in the eastern Denver Basin, Colorado. *Rocky Mountain Geology*, **38**, 45–71. https://doi.org/10.2113/gsrocky.38.1.45.

Barthelmes, M. (2022). *Pinnacles National Park: Geologic Resources Inventory Report*. Fort Collins, CO: National Park Service. Natural Resource Report NPS/NRSS/GRD/NRR—2022/2471. https://doi.org/10.36967/2294842.

Beard, L. S. (1996). Paleogeography of the Horse Spring Formation in relation to the Lake Mead fault system, Virgin Mountains, Nevada and Arizona.

Geological Society of America Special Paper, **303**, 27–60. https://doi.org/10.1130/0-8137-2303-5.27.

Belt, E. S., Hartman, J. H., Diemer, J. A., et al. (2004). Unconformities and age relationships, Tongue River and older members of the Fort Union Formation (Paleocene), western Williston Basin, U. S. A. *Rocky Mountain Geology*, **39**(2), 113–140. https://doi.org/10.2113/39.2.113.

Benedict, J. C., DeVore, M. L., & Pigg, K. B. (2011). *Prunus* and *Oemleria* (Rosaceae) flowers from the late Early Eocene Republic Flora of northeastern Washington State, U.S.A. *International Journal of Plant Sciences*, **172**(7), 948–958.

Bennett, M. R., Bustos, D., Pigati, J. S., et al. (2021a). Evidence of humans in North America during the Last Glacial Maximum. *Science*, **373**(6562), 1528–1531. https://doi.org/10.1126/science.abg7586.

Bennett, M. R., Bustos, D., Pigati, J. S., et al. (2021b). Supplementary materials for "Evidence of humans in North America during the Last Glacial Maximum." *Science*, **373**(6562), 1528–1531. https://doi.org/10.1126/science.abg7586.

Berry, E. W. (1906). Pleistocene plants from Virginia. *Torreya*, **6**, 88–90.

Berry, E. W. (1909). Pleistocene swamp deposits in Virginia. *The American Naturalist*, **43**, 432–436.

Berry, E. W. (1911). Flora of the Raritan Formation. *Geological Survey of New Jersey Bulletin*, **3**, 1–233.

Berry, E. W. (1916a). The Lower Eocene floras of southeastern North America. *U.S. Geological Survey Professional Paper*, **91**, 1–481. https://doi.org/10.3133/pp91.

Berry, E. W. (1916b). The physical conditions indicated by the flora of the Calvert Formation. *U.S. Geological Survey Professional Paper*, **98-F**, 61–73. https://doi.org/10.3133/pp98F.

Berry, E. W. (1930). A new Miocene *Cercis* from Idaho and Washington. *Bulletin of the Torrey Botanical Club*, **57**(4), 239–244. https://doi.org/10.2307/2480616.

Berry, E. W. (1932). The Miocene flora of Idaho. *Proceedings of the National Academy of Sciences*, **18**(4), 289–292. https://doi.org/10.1073/pnas.18.4.289.

Berry, E. W. (1933). New occurrence of Pleistocene plants in the District of Columbia. *Journal of Washington Academy of Science*, **23**(1), 1–25.

Berry, E. W. (1934). Miocene plants from Idaho. *U.S. Geological Survey Professional Paper*, **185-E**, 97–125. https://doi.org/10.3133/pp185E.

Berry, E. W. (1938). Pleistocene fossils from Westmoreland County, Virginia. *Journal of the Washington Academy of Sciences*, **28**, 58–61.

Bestland, E. A., & Retallack, G. J. (1994). *Geology and Paleoenvironments of the Clarno Unit, John Day Fossil Beds National Monument, Oregon*, National Park Service contract CX-9000-1-10009. Kimberly, OR: John Day Fossil Beds National Monument.

Beyer, A. F. (1954). Some petrified wood from the Specimen Ridge area of Yellowstone National Park. *American Midland Naturalist*, **51**(2), 553–576. https://doi.org/10.2307/2422124.

Biek, R. F., Anderson, J. J., Rowley, P. D., & Maldonado, F. (2011). Interim geologic map of the west part of the Panguitch 30′ x 60′ quadrangle, Garfield, Iron, and Kane Counties, Utah. *Utah Geological Survey Open-File Report*, **585**, 1 map, scale 1:62,500.

Biek, R. F., & Gonzalez, M. A. (2001). The geology of Theodore Roosevelt National Park: Billings and McKenzie counties, North Dakota. *North Dakota Geological Survey Miscellaneous Series*, **86**, 1–74.

Black D. M. [1962]. Preliminary Inventory of Geological Features of Natchez Trace Parkway. Natchez Trace Parkway files. NPS TIC#D-551; Kenworthy et al., 2007.

Blair, W. N., & Armstrong, A. K. (1979). Hualapai Limestone Member of the Muddy Creek Formation: The youngest deposit predating the Grand Canyon, southeastern Nevada, and northwestern Arizona. *U.S. Geological Survey Professional Paper*, **1111**, 1–14. https://doi.org/10.3133/pp1111.

Blodgett, R. B., Santucci, V. L., & Tweet J. S. (2016). An inventory of paleontological resources from Katmai National Park and Preserve, southwest Alaska. *New Mexico Museum of Natural History and Science Bulletin*, **74**, 41–50.

Blonder, B., Royer, D. L., Johnson, K. R., Miller, I., & Enquist, B. J. (2014). Plant ecological strategies shift across the Cretaceous–Paleogene boundary. *PLoS Biology*, **12**(9), e1001949. https://doi.org/10.1371/journal.pbio.1001949.

Bohannon, R. G. (1984). Nonmarine sedimentary rocks of Tertiary age in the Lake Mead Region, southeastern Nevada and northwestern Arizona. *U.S. Geological Survey Professional Paper*, **1259**, 1–72. https://doi.org/10.3133/pp1259.

Bonde, A., Santucci, V. L., Tweet, J. S., Eichenberg, E., & Moore, B. (2018). *Lake Mead National Recreation Area: Paleontological Resources Inventory*. Fort Collins, CO: National Park Service. Natural Resource Report NPS/LAKE/NRR—2018/1618.

Bonnichsen, B., & Breckenridge, R. M., eds. (1982). *Cenozoic Geology of Idaho*. Moscow, ID: Idaho Department of Lands, Bureau of Mines and Geology.

Bouchal, J. M., Zetter, R., & Denk, T. (2016). Pollen and spores of the uppermost Eocene Florissant Formation, Colorado: A combined light and scanning electron microscopy study. *Grana*, **55**(3), 179–245. https://doi.org/10.1080/00173134.2015.1108362.

Britton, E. G., & Hollick, A. (1907). American fossil mosses, with description of a new species from Florissant, Colorado. *Bulletin of the Torrey Botanical Club*, **34**, 139–149. https://doi.org/10.2307/2479149.

Britton, E. G., & Hollick, A. (1915). A new American fossil moss. *Bulletin on the Torrey Botanical Club*, **42**, 9–10. https://doi.org/10.2307/2479420.

Broecker, W. S., & Farrand., W. R. (1963). Radiocarbon age of the Two Creeks Forest Bed, Wisconsin. *Geological Society of America Bulletin*, **74**(6), 795–802. https://doi.org/10.1130/0016-7606(1963)74[795:RAOTTC]2.0.CO;2.

Brown, R. A. (2014). *Description of the High-Altitude Centennial Fossil Site in Wind Cave National Park, Custer County, South Dakota*. Rapid City, SD: South Dakota School of Mines and Technology, master's thesis.

Brown, R. W. (1935). Miocene leaves, fruits, and seeds from Idaho, Oregon, and Washington. *Journal of Paleontology*, **9**(7), 572–587.

Brown, R. W. (1937). Additions to some fossil floras of the western United States. *U.S. Geological Survey Professional Paper*, **186-J**, 163–206. https://doi.org/10.3133/pp186J.

Brown, R. W. (1956). New items in Cretaceous and Tertiary floras of the western United States. *Journal of the Washington Academy of Sciences*, **46**(4), 104–108.

Browning, J. V., Sugarman, P. J., Miller, K. G., et al. (2011). Double Trouble site. In Miller, K. G., Sugarman, P. J., Browning, K. G., et al., *Proceedings of the Ocean Drilling Program*, **174AX** (supplement), 1–63.

Burnham, R. J. (1994). Paleoecological and floristic heterogeneity in the plant-fossil record: An analysis based on the Eocene of Washington. *U.S. Geological Survey Bulletin*, **2085-B**, B1–B36. https://doi.org/10.3133/b2085B.

Calderwood, K. W., & Fackler, W. C. (1972). Proposed stratigraphic nomenclature for Kenai Group, Cook Inlet Basin, Alaska. *American Association of Petroleum Geologists Bulletin*, **56**(4), 739–754. https://doi.org/10.1306/819A407E-16C5-11D7-8645000102C1865D.

Carpenter, S. M., Bousman, B. C., Potapova, O., et al. (2013). The San Antonio River Mammoth Site: Archaeological testing investigations for the Interstate 37 Bridge at the San Antonio River improvement project, Bexar County, Texas. *Index of Texas Archaeology: Open Access Gray Literature from the Lone Star State*, **2013**(3), 1–208. https://doi.org/10.21112/ita.2013.1.3.

Carr, M. S. (1950). The District of Columbia, its rocks and their geologic history. *U.S. Geological Survey Bulletin*, **967**, 1–59. https://doi.org/10.3133/b967.

Carvalho, M. R., Jarmillo, C., de la Parra, F., et al. (2021). Extinction at the end-Cretaceous and the origin of modern Neotropical rainforests. *Science*, **372**(6537), 63–68. https://doi.org/10.1126/science.abf1969.

Cenozoic CO Proxy Integration Project (CenCOPIP) Consortium. (2023). Toward a Cenozoic history of atmospheric CO_2. *Science*, **382**(6675), eadi5177. https://doi.org/10.1126/science.adi5177.

Chadwick, A., & Yamamoto, T. (1984). A paleoecological analysis of the petrified trees in the Specimen Creek area of Yellowstone National Park, Montana, USA. *Palaeogeography, Palaeoclimatology, Palaeoecology*, **45**(1), 39–48. https://doi.org/10.1016/0031-0182(84)90108-1.

Chamberlain, J. A., Jr., Stoffer, P. W., & Messina, P. (2006). Fossiliferous nodules from New York Gateway beaches. *New Mexico Museum of Natural History and Science Bulletin*, **34**, 11.

Chaney, R. W. (1925). The Mascall Flora: Its distribution and climatic relation. *Carnegie Institution of Washington Publication*, **349**, 23–48.

Chaney, R. W. (1927). Geology and paleontology of the Crooked River Basin, with special reference to the Bridge Creek flora. *Carnegie Institute of Washington Publications*, **346**, 45–138.

Chaney, R. W., & Axelrod, D. I. (1959). Miocene floras of the Columbian Plateau. *Carnegie Institution Washington Publication*, **617**, 1–237.

Chaney, R. W., & Mason, H. L. (1930). A Pleistocene flora from Santa Cruz Island, California. *Carnegie Institution of Washington Publication*, **415**, 1–24.

Clark, F. E. (1983). Benthonic foraminifera assemblages, basal Modelo Formation, Big Mountain, northern Simi Valley, California. In Squires, R. L., & Filewicz, M. V., eds., *Cenozoic Geology of the Simi Valley Area*. Los Angeles: Society of Economic Paleontologists and Mineralogists, Pacific Section, pp. 203–205.

Clark, J. (1967). General geology. *Fieldiana: Geology Memoirs*, **5**, 9–15.

Clinton, J. M., Visaggi, C. C., Tweet, J. S., Santucci, V. L., & Irick, K. (2023). *Gulf Islands National Seashore: Paleontological Resource Inventory (Public Version)*. Fort Collins, CO: National Park Service. Natural Resource Report NPS/GUIS/NRR—2023/2525. https://doi.org/10.36967/2299352.

Cockerell, T. D. A. (1908). Descriptions of tertiary plants (I). *American Journal of Science, series 4*, **26**, 65–68.

Cockerell, T. D. A. (1915). *Equisetum* in the Florissant Miocene. *Torreya*, **15**, 265–267.

Cockerell, T. D. A. (1938). San Miguel Island, California. *The Scientific Monthly*, **46**(2), 180–187.

Coffin, H. G. (1984). Preliminary report: Identification of petrified wood from Theodore Roosevelt National Park. Internal report to Theodore Roosevelt National Park.

Cohen, K. M., Finney, S. C., Gibbard, P. L., & Fan, J.-X. (2013; updated 10/2022). The ICS International Chronostratigraphic Chart. *Episodes* **36**, 199–204. https://doi.org/10.18814/epiiugs/2013/v36i3/002.

Colburn, I. P., Jakobsen, C. M., & Novak, G. A. (1988). The Paleocene stratigraphy of the Santa Monica Mountains, Los Angeles County, California. In Filewicz, M. V., & Squires, R. L., eds., *Paleogene Stratigraphy, West Coast of North America*. Los Angeles: Society of Economic Paleontologists and Mineralogists, Pacific Section, pp. 59–72.

Collinson, M. E. (2001). Cainozoic ferns and their distribution. *Brittonia*, **53**, 173–235.

Collinson, M. E. (2002). The ecology of Cainozoic ferns. *Review of Palaeobotany and Palynology*, **119**, 51–68. https://doi.org/10.1016/S0034-6667(01)00129-4.

Constenius, K. N., Dawson, M. R., Pierce, H. G., Walter, R. C., & Wilson, M. V. (1989). Reconnaissance paleontologic study of the Kishenehn Formation, northwestern Montana and southeastern British Columbia. In French, D. E., & Grabb, R. F., eds., *1989 field Conference Guidebook: Montana Centennial Edition: Geologic Resources of Montana, Volume 1*. Billings, MT. Montana Geological Society, pp. 189–203.

Crane, P. R. (1989). Paleobotanical evidence on the early radiation of nonmagnoliid dicotyledons. *Plant Systematics and Evolution*, **162**(1/4), 165–191. https://doi.org/10.1007/BF00936916.

Crane, P. R., Manchester, S. R., & Dilcher, D. L. (1990). A preliminary survey of fossil leaves and well-preserved reproductive structures from the Sentinel Butte Formation (Paleocene) near Almont, North Dakota. *Fieldiana: Geology*, New Series **20**(1418). https://doi.org/10.5962/bhl.title.100826.

Crane, P. R., Manchester, S. R., & Dilcher, D. L. (1991). Reproductive and vegetative structure of *Nordenskioldia* (Trochodendraceae), a vesselless dicotyledon from the Early Tertiary of the Northern Hemisphere. *American Journal of Botany*, **78**(10), 1311–1334. https://doi.org/10.1002/j.1537-2197.1991.tb12599.x.

Crepet, W. L., & Nixon, K. C. (1989). Earliest megafossil evidence of Fagaceae: Phylogenetic and biogeographic implications. *American Journal of Botany*, **76**(6), 842–855. https://doi.org/10.1002/j.1537-2197.1989.tb15062.x.

Crepet, W. L., Nixon, K. C., & Gandolfo, M. A. (2001). Turonian flora of New Jersey, USA. *Asociacion Paleontologica Argentina. Publicacion Especial*, 7, 61–69.

Criswell R. [1968]. A Paleontological Study of the Gold Bluffs Area of Prairie Creek Redwoods State Park, California. Arcata, CA: Humboldt State University; Trexler, 1989.

Csejtey, B., Jr., Mullen, M. W., Cox, D. P., et al. (1986). Geology and geochronology of the Healy Quadrangle, Alaska. *U.S. Geological Survey Open-File Report*, **86-396**, 1–92. https://doi.org/10.3133/ofr86396.

Csejtey, B., Jr., Mullen, M. W., Cox, D. P., & Stricker, G. D. (1992). Geology and geochronology of the Healy quadrangle, south-central Alaska. *U.S. Geological Survey Miscellaneous Investigation Series Map*, **1961**, 2 sheets, scale 1:250,000. https://doi.org/10.3133/i1961.

Culberson, W. L. (1955). The fossil mosses of the Two Creeks Forest Bed of Wisconsin. *The American Midland Naturalist*, **54**(2), 452–459. https://doi.org/10.2307/2422580.

Danehy, D. R., Wilf, P., & Little, S. A. (2007). Early Eocene macroflora from the Red Hot Truck Stop locality (Meridian, Mississippi, USA). *Palaeontologia Electronica*, **10**(3, article 17A), 1–31.

Darton, N. H. (1950). Bedrock subsurface of the District of Columbia and vicinity. *U.S. Geological Survey Professional Paper*, **217**, 1–42. https://doi.org/10.3133/pp217.

Davis, M., Douglas, C., Calcote, R., et al. (2000). Holocene climate in the western Great Lakes National Parks and Lakeshores: implications for future climate change. *Conservation Biology*, **14**(4), 968–983. https://doi.org/10.1046/j.1523-1739.2000.99219.x.

Davis, O. K., & Shafer, D. S. (1992). An early-Holocene maximum for the Arizona monsoon recorded at Montezuma Well, central Arizona. *Palaeogeography, Palaeoclimatology, Palaeoecology*, **92**, 107–119. https://doi.org/10.1016/0031-0182(92)90137-T.

Davis, O. K., & Shafer, D. S. (2006). *Sporormiella* fungal spores, a palynological means of detecting herbivore density. *Palaeogeography, Palaeoclimatology, Palaeoecology*, **237**(1), 40–50. https://doi.org/10.1016/j.palaeo.2005.11.028.

Dawson, M. R., & Constenius, K. N. (2018). Mammalian fauna of the middle Eocene Kishenehn Formation, middle fork of the Flathead River, Montana. *Annals of Carnegie Museum*, **85**(1), 25–60. https://doi.org/10.2992/007.085.0103.

Denk, T., Grímsson, F., Zetter, R., & Símonarson, L. A. (2011). The biogeographic history of Iceland – the North Atlantic Land Bridge revisited. In

Landman, N. H., & Harries, P., eds., *Late Cainozoic Floras of Iceland*. Dordrecht: Kluwer Academic/Plenum Publishers, Topics in Geobiology, **35**, 647–668. https://doi.org/10.1007/978-94-007-0372-8_12.

Detterman, R. L., Case, J. E., Miller, J. W., Wilson, F. H., & Yount, M. E. (1996). Stratigraphic framework of the Alaska Peninsula. *U.S. Geological Survey Bulletin*, **1969-A**, 1–74. https://doi.org/10.3133/b1969A.

Detterman, R. L., Miller, T. P., Yount, M. E., & Wilson, F. H. (1981). Geologic map of the Chignik and Sutwik Island Quadrangles, Alaska. *U.S. Geological Survey Miscellaneous Investigations Series Map*, **1229**, 1 sheet, scale 1:250,000. https://doi.org/10.3133/i1229.

Deussen, A. (1924). Geology of the Coastal Plain of Texas west of Brazos River. *U.S. Geological Survey Professional Paper*, **126**, 1–139. https://doi.org/10.3133/pp126.

DeVore, M. L., & Pigg, K. B. (2007). A brief review of the fossil history of the family Rosaceae with a focus on the Eocene Okanogan Highlands of eastern Washington State, USA, and British Columbia, Canada. *Plant Systematics and Evolution*, **266**(1–2), 45–57. https://doi.org/10.1007/s00606-007-0540-3.

DeVore, M. L., & Pigg, K. B. (2010). Floristic composition and comparison of middle Eocene to late Eocene and Oligocene floras in North America. *Bulletin of Geosciences*, **85**(1), 111–134.

DiBenedetto, J. N. (1998). An Oligocene tree stump from Badlands National Park, scenic member of the Brule Formation, Interior, South Dakota. *Conference on Fossil Resources Abstracts with Programs*, **5**.

Diffendal, R. F. Jr., Pabian, R. K., & Thomasson, J. R. (1996). Geologic history of Ash Hollow State Historical Park, Nebraska. *Conservation and Survey Division Educational Circular*, **5A**, 1–36.

Dilcher, D. (1971). Revision of the Eocene flora of southeastern North America. *Journal of Palaeosciences*, **20**(1–3), 7–18. https://doi.org/10.54991/jop.1971.882.

Dillhoff, R. M., Dillhoff, T. A., Dunn, R. E., Myers, J. A., & Strömberg, C. A. E. (2009). Cenozoic paleobotany of the John Day Basin, central Oregon. *Geological Society of America Field Guide*, **15**, 135–164. https://doi.org/10.1130/2009.fld015(07).

Dorf, E. (1933). Pliocene floras of California. *Carnegie Institute of Washington Publication*, **412**, 1–112.

Dorf, E. (1960). Tertiary fossil forests of Yellowstone National Park, Wyoming. *Billings Geological Society Guidebook*, **11**, 253–260.

Dorf, E. (1964). The petrified forests of Yellowstone Park. *Scientific American*, **210**(4), 106–115.

Dover, J. H., & Miyaoka, R. T. (1988). Reinterpreted geologic map and fossil data, Charley River Quadrangle, east-central Alaska. *U.S. Geological Survey Miscellaneous Field Studies Map*, **2004**, 2 sheets, 1:250,000. https://doi.org/10.3133/mf2004.

Dunwiddie, P. W. (1983). *Holocene Forest Dynamics on Mount Rainier, Washington*. Seattle: University of Washington, doctoral dissertation.

Dunwiddie, P. W. (1985). Dichotomous key to conifer foliage in the Pacific Northwest. *Northwest Science*, **59**(3), 185.

Dunwiddie, P. W. (1986). A 6,000-year record of forest history on Mount Rainier, Washington. *Ecology*, **67**(1), 58–68. https://doi.org/10.2307/1938503.

Dunwiddie, P. W. (1987). Macrofossil and pollen representation of coniferous trees in modern sediments from Washington. *Ecology*, **68**(1), 1–11. https://doi.org/10.2307/1938800.

Eaton, J. G., Tibert, N. E., & Biek, R. F. (2011). First mammals and ostracodes from the Paleogene Claron Formation, southwestern Utah. *Geological Society of America Abstracts with Programs*, **43**(4), 77.

Edwards, E. J., Osborne, C. P., Strömberg, C. A. E., et al. (2010). The origins of C4 grasslands: Integrating evolutionary and ecosystem science. *Science*, **328** (5978), 587–591. https://doi.org/10.1126/science.1177216.

Elder, W. P., Nyborg, T., Kenworthy, J. P., & Santucci, V. L. (2008). *Paleontological Resource Inventory and Monitoring: San Francisco Bay Area Network*. Fort Collins, CO: National Park Service. Natural Resource Technical Report NPS/NRPC/NRTR—2008/078.

Elder, W. P., Santucci, V. L., Kenworthy, J. P., Blodgett, R. B., & McKenna, R. T. P. (2009). *Paleontological Resource Inventory and Monitoring: Arctic Network*. Fort Collins, CO: National Park Service. Natural Resource Technical Report NPS/NRPC/NRTR— 2009/276.

Elias, S. A., Kuzmina, S., & Kiselyov, S. (2006). Late Tertiary origins of the Arctic beetle fauna. *Palaeogeography, Palaeoclimatology, Palaeoecology*, **241**(3–4), 373–392.

Epis, C. R., & Chapin, C. E. (1975). Geomorphic and tectonic implications of the post-Laramide, late Eocene erosion surface in the southern Rocky Mountains. *Geological Society of America Memoir*, **144**, 45–74. https://doi.org/10.1130/MEM144-p45.

Erdei, B., & Manchester, S. R. (2015). Ctenis clarnoensis sp. n., an unusual cycadalean foliage from the Eocene Clarno Formation, Oregon. *International Journal of Plant Sciences*, **176** (1), 31–43. https://doi.org/10.1086/678467.

Erickson, B. R. (1999). Fossil Lake Wannagan (Paleocene: Tiffanian): Billings County, North Dakota. *North Dakota Geological Survey Miscellaneous Series*, **87**, 1–9.

Evanoff, E., McIntosh, W. C., & Murphey, P. C. (2001). Stratigraphic summary and ^{40}Ar/^{39}Ar geochronology of the Florissant Formation, Colorado. *Proceedings of the Denver Museum of Nature and Science, Series 4*, **1**, 1–16.

Evans, J. E. (1998). Facies associations of freshwater carbonates in the Eocene-Oligocene Chadron Formation, White River Group, Nebraska and South Dakota, USA. *Utah Geological Association Publication*, **26**, 209–231.

Evans, J. E. (1999). Recognition and implications of Eocene tufas and travertines in the Chadron Formation, White River Group, Badlands of South Dakota. *Sedimentology*, **46**(5), 771–789. https://doi.org/10.1046/j.1365-3091.1999.00250.x.

Fastovsky, D. E., & McSweeney, K. (1991). Paleocene paleosols of the petrified forests of Theodore Roosevelt National Park, North Dakota: A natural experiment in compound pedogenesis. *Palaios*, **6**(1), 67–80. https://doi.org/10.2307/3514954.

Fay, C. L., Kenworthy, J. P., & Santucci, V. L. (2009). *Paleontological Resources Inventory and Monitoring: North Coast and Cascades Network*. Fort Collins, CO: National Park Service. Natural Resources Technical Report NPS/NRPC/NRTR—2009/250.

Fergusson, G. J., & Libby, W. F. (1963). UCLA radiocarbon dates II. *Radiocarbon* 5, 1–22. https://doi.org/10.1017/S0033822200036742.

Fiorillo, A. R., Fanti, F., Hults, C., & Hasiotis, S. T. (2014). New ichnological, paleobotanical, and detrital zircon data from an unnamed rock unit in Yukon–Charley Rivers National Preserve (Cretaceous: Alaska): Stratigraphic implications for the region. *Palaios*, **29**(1), 16–26. https://doi.org/10.2110/palo.2013.054.

Fisher, R. V. (1957). *Stratigraphy of the Puget Group and Keechelus Group in the Elbe-Packwood Area of South-Western Washington*. Seattle: University of Washington, doctoral dissertation.

Fisher, T. G., & Whitman, R. L. (1999). Deglacial and lake level fluctuation history recorded in cores, Beaver Lake, Upper Peninsula, Michigan. *Journal of Great Lakes Research*, **25**(2), 263–274. https://doi.org/10.1016/S0380-1330(99)70735-5.

Fiske, R. S., Hopson, C. A., & Waters, A. C. (1963). Geology of Mount Rainier National Park, Washington. *U.S. Geological Survey Professional Paper*, **444**, 1–93. https://doi.org/10.3133/pp444.

Flack, M. E. (1993). Depositional environments of the "Topanga Canyon" Formation, the Encinal Tongue of the Rincon Shale, and the Fernwood

Tongue of the Sespe Formation, central Santa Monica Mountains, California. *Pacific Section SEPM (Society for Sedimentary Geology) Book*, **72**, 45–74.

Florer, L. E. (1972). Palynology of a postglacial bog in the New Jersey pine barrens. *Bulletin of the Torrey Botanical Club*, **99**(3), 135–138. https://doi.org/10.2307/2484694.

Flynn, A. G., & Peppe, D. J. (2019). Early Paleocene tropical forest from the Ojo Alamo Sandstone, San Juan Basin, New Mexico, USA. *Paleobiology*, **45**(4), 612–635. https://doi.org/10.1017/pab.2019.24.

Fontaine, W. M. (1882). The artesian well at Fort Monroe, Va. *The Virginias*, **3**(2), 18–19.

Forest, F., Savolainen, V., Chase, M. W., et al. (2005). Teasing apart molecular- versus fossil-based error estimates when dating phylogenetic trees: A case study in the birch family (Betulaceae). *Systematic Botany*, **30**(1), 118–133. https://doi.org/10.1600/0363644053661850.

Forester, R. (1979). Report on referred fossils [Unnamed formation, Survey Pass quadrangle, Sample No. 79Aha081, 11/19/1979]. Alaska Paleontological Database, www.alaskafossil.org/.

Foster, H. L. (1976). Geologic map of the Eagle quadrangle, Alaska. *U.S. Geological Survey Miscellaneous Investigations Series Map*, **922**, 1 sheet, scale 1:250,000. https://doi.org/10.3133/i922.

Fredlund, G. G., & Tieszen, L. L. (1997). Phytolith and carbon isotope evidence for late Quaternary vegetation and climate change in the southern Black Hills, South Dakota. *Quaternary Research*, **47**(2), 206–217. https://doi.org/10.1006/qres.1996.1862.

Fremd, T. J., Bestland, E. A., & Retallack, G. J. (1994). *John Day Basin Field Trip Guide and Road Log.* Seattle, WA: Northwest Interpretive Association. 1994 Society of Vertebrate Paleontology field trip guidebook.

Fremd, T. J., Dunn, R. E., Rickabaugh, S. J., Graham, D. B., & Rosenkrans, D. (2003). Paleontological reconnaissance of the Frederika Formation, Wrangell–St. Elias National Park and Preserve, southeastern Alaska. *Geological Society of America Abstracts with Programs*, **35**(6), 500.

Fritz, W. J. (1980). Reinterpretation of the depositional environment of the Yellowstone "fossil forests." *Geology*, **8**(7), 309–313. https://doi.org/10.1130/0091-7613(1980)8%3C309:ROTDEO%3E2.0.CO;2.

Fritz, W. J. (1982). Geology of the Lamar River Formation, northeast Yellowstone National Park. *Wyoming Geological Association Guidebook*, **33**, 73–101.

Fritz, W. J. (1986). Plant taphonomy in areas of explosive volcanism in land plants. *Studies in Geology*, **15**, 1–9. https://doi.org/10.1017/S0271164800001299.

Fritz, W. J. (1994). *Report on the Lake Butte Fossil Locality, Yellowstone National Park*. Internal report to Yellowstone National Park, MT.

Fritz, W. J., & Fisk, L. H. (1978). Eocene petrified woods from one unit of the Amethyst Mountain "Fossil Forest." *Northwest Geology*, **7**, 10–19.

Gabel, M. L., Backlund, D. C., & Haffner, J. (1998). The Miocene macroflora of the northern Ogallala Group, northern Nebraska and southern South Dakota. *Journal of Paleontology*, **72**(2), 388–397. https://doi.org/10.1017/S0022336000036362.

Galloway, A. J. (1977). Geology of the Point Reyes Peninsula, Marin County, California. *California Division of Mines and Geology Bulletin*, **202**, 1–72.

Goetcheus, V. G., & Birks, H. H. (2001). Full-glacial upland tundra vegetation preserved under tephra in the Beringia National Park, Seward Peninsula, Alaska. *Quaternary Science Reviews*, **20**(1–3), 135–147. https://doi.org/10.1016/S0277-3791(00)00127-X.

Gohn, G. S., Brewster-Wingard, G. L., Cronin, T. M., et al. (1996). Neogene and Quaternary geology of a stratigraphic test hole on Horn Island, Mississippi Sound. *U.S. Geological Survey Open-File Report*, **96-20A**, 1–23. https://doi.org/10.3133/ofr9620A.

Golenberg, E., Giannasi, D., Clegg, M., et al. (1990). Chloroplast DNA sequence from a Miocene *Magnolia* species. *Nature*, **344**(6267), 656–658. https://doi.org/10.1038/344656a0.

Golovneva, L. B., & Alekseev, P. I. (2017). Taxonomy and morphological diversity of infructescences *Jenkinsella* co-occurred with *Trochodendroides* leaves in the Cretaceous and Paleogene. *Палеоботаника*, **8**, 92–121. http://dx.doi.org/10.31111/palaeobotany/2017.8.92.

Graham, A. (1965). Outline of the origin and historical affinities between Asia and eastern North America. In Graham, A. ed., *Floristics and Paleofloristics of Asia and Eastern North America*. Amsterdam: Elsevier Publishing Co., pp. 201–233.

Graham, A. (1999). *Late Cretaceous and Cenozoic History of North American Vegetation North of Mexico*. New York: Oxford University Press.

Graham, J. P. (2009). *Scotts Bluff National Monument Geologic Resources Inventory Report*. Denver, CO: National Park Service. Natural Resource Report NPS/NRPC/GRD/NRR—2009/085.

Graham, J. P. (2011). *Geologic Resources Inventory Scoping Summary: Herbert Hoover National Historic Site, Iowa*. Fort Collins, CO: National Park Service.

Graham, J. P. (2017). *Herbert Hoover National Historic Site: Geologic Resources Inventory Report*. Fort Collins, CO: National Park Service. Natural Resource Report NPS/NRSS/GRD/NRR—2017/1479.

Grande, L. (2013). *Snapshots from Deep Time: The Lost World of Fossil Lake*. Chicago: University of Chicago Press.

Grandstaff, D. E., & Terry Jr., D. O. (2009). Rare earth element composition of Paleogene vertebrate fossils from Toadstool Geologic Park, Nebraska, USA. *Applied Geochemistry* **24**(4), 733–745. https://doi.org/10.1016/j.apgeochem.2008.12.027.

Greenwalt, D. E., Rose, T. R., Siljestrom, S. M., et al. (2014). Taphonomy of the fossil insects of the middle Eocene Kishenehn Formation. *Acta Palaeontologica Polonica*, **60**(4), 931–947. https://doi.org/10.4202/app.00071.2014.

Gregory, H. E. (1950). Geology of eastern Iron County, Utah. *Utah Geological and Mineralogical Survey Bulletin*, **37**, 1–153.

Grissino Mayer, H. D., & Swetnam, T. W. (2000). Century scale climate forcing of fire regimes in the American Southwest. *Holocene*, **10** (2), 213–220. https://doi.org/10.1191/095968300668451235.

Grove, K., Colson, K., Binkin., M., Dull, R., & Garrison, C. (1995). Stratigraphy and structure of the Late Pleistocene Olema Creek Formation, San Andreas fault zone north of San Francisco, California. *Pacific Section SEPM (Society for Sedimentary Geology) Book*, **76**, 55–77.

Haire, S. A., Hanks, H. D., Tremain, E. S., et al. (2001). Fossil insect damage on Paleocene leaves from Wannagan Creek, ND. *Geological Society of America Abstracts With Programs* **33**(6), 197.

Hamilton, D. L. (1964). Hawaii's trees of stone. *National Parks Magazine*, **January 1964**, 15–17.

Hamilton, T. D., & Ashley, G. M. (1993). Epiguruk: A late Quaternary environmental record from northwestern Alaska. *Geological Society of America Bulletin*, **105**(5), 583–602. https://doi.org/10.1130/0016-7606(1993)105%3C0583:EALQER%3E2.3.CO;2.

Hamilton, T. D., Ashley, G. M., Reed, K. M., & Schweger, C. E. (1993). Late Pleistocene vertebrates and other fossils from Epiguruk, northwestern Alaska. *Quaternary Research*, **39**, 381–387. https://doi.org/10.1006/qres.1993.1045.

Hamilton, T. D., Ashley, G. M., Reed, K. M., & van Etten, D. P. (1984). Stratigraphy and sedimentology of Epiguruk Bluff: A preliminary account. *U.S. Geological Survey Circular*, **939**, 12–15. https://doi.org/10.3133/cir939.

Han, M., Manchester, S. R., Fu, Q. Y., Jin, J. H., & Quan, C. (2018). Paleogene fossil fruits of *Stephania* (Menispermaceae) from North America and East Asia. *Journal of Systematics and Evolution*, **56**(2), 81–91. https://doi.org/10.1111/jse.12288.

Hanson, C. B. (1996). Stratigraphy and vertebrate faunas of the Bridgerian-Duchesnean Clarno Formation, north-central Oregon. In Prothero, D. R., & Emry, R. J., eds., *The Terrestrial Eocene–Oligocene Transition in North America*. New York: Cambridge University Press, pp. 206–239.

Harr, J. L., & Ting, F. T. (1977). Modern and Paleocene *Metasequoias*: A comparison of foliage morphology. In Romans, R. C., ed., *Geobotany*. Boston, MA: Springer, pp. 109–131. https://doi.org/10.1007/978-1-4757-1674-0_7.

Harrington, G. J. (2008). Comparisons between Palaeocene–Eocene paratropical swamp and marginal marine pollen floras from Alabama and Mississippi, USA. *Palaeontology*, **51**(3), 611–622. https://doi.org/10.1111/j.1475-4983.2008.00768.x.

Harrington, G. J., Kemp, S. J., & Koch, P. L. (2004). Palaeocene–Eocene paratropical floral change in North America: Responses to climate change and plant immigration. *Journal of the Geological Society*, **161**, 173–184. https://doi.org/10.1144/0016-764903-100.

Harrington, M. D., Lanik, A., Hults, C. P., & Druckenmiller, P. S. (2023). *Focused Condition Assessment of Paleontological Resources Within Katmai National Park and Preserve*. Fort Collins, CO: National Park Service. Natural Resource Report NPS/KATM/NRR—2023/2565. https://doi.org/10.36967/2298782.

Hazzard, J. C. (1954). Rocks and structure of the northern Providence Mountains, San Bernardino County, California. *California Division of Mines Bulletin*, **170**(1 V), 27–36.

Heer, O. (1869). *Flora Fossilis Alaskana*. Stockholm: P. A. Norstedt & Söner. Kongl. Svenska Vetenskaps Akademiens Handlinger, **8**(4), 1–41.

Henkel, C. J., Elder, W. P., Santucci, V. L., & Clites, E. C. (2015). *Golden Gate National Recreation Area Paleontological Resource Inventory*. Fort Collins, CO: National Park Service. Natural Resource Report NPS/GOGA/NRR—2015/915.

Herendeen, P. S., Cardoso, D. B. O. S., Herrera, F., & Wing, S. L. (2022). Fossil papilionoids of the *Bowdichia* clade (Leguminosae) from the Paleogene of North America. *American Journal of Botany*, **109**(1), 130–150. https://doi.org/10.1002/ajb2.1808.

Herendeen, P. S., & Dilcher, D. L. (1991). *Caesalpinia* subgenus *Mezoneuron* (Leguminosae, Caesalpinioideae) from the Tertiary of North America. *American Journal of Botany*, **78**, 1–12. https://doi.org/10.1002/j.1537-2197.1991.tb12566.x.

Hermsen, E. J. (2021). Review of the fossil record of *Passiflora*, with a description of new seeds from the Pliocene Gray Fossil Site, Tennessee,

USA. *International Journal of Plant Sciences*, **182**(6), 533–550. https://doi.org/10.1086/714282.

Hermsen, E. J. (2023). Pliocene seeds of *Passiflora* subgenus *Decaloba* (Gray Fossil Site, Tennessee) and the impact of the fossil record on understanding the diversification and biogeography of *Passiflora*. *American Journal of Botany*, **110**(3), e16137. https://doi.org/10.1002/ajb2.16137.

Herring, T., Tweet, J. S., & Santucci, V. L. (2023). *Wind Cave National Park: Paleontological Resource Inventory (Public Version)*. Fort Collins, CO: National Park Service. Natural Resource Report NPS/WICA/NRR—2023/2536. https://doi.org/10.36967/2299620.

Hevly, R. H. (1974). Recent paleoenvironments and geological history at Montezuma Well. *Journal of the Arizona Academy of Science*, **9**(2), 66–75. https://doi.org/10.2307/40025632.

Hevly, R. H., Nations, J. D., & Zabo, B. (1992). History and environment of Pleistocene tufa deposition at Montezuma Well and Red Tank Draw, Verde Valley, Arizona. In *Program and Abstracts, Proceedings of the Southwestern and Rocky Mountain Division, American Association for the Advancement of Science, Sixty-Eighth Annual Meeting*, abstract 125, 28.

Hickey, L. J. (1977). Stratigraphy and paleobotany of the Golden Valley Formation (early Tertiary) of western North Dakota. *Geological Society of America Memoir*, **150**, 1–183. https://doi.org/10.1130/MEM150.

Hodgkinson, K. A. (1974). Stone City and Cook Mountain (middle Eocene) scaphopods from southwest Texas. *The University of Kansas Paleontological Contributions*, **70**, 1–25.

Hodnett, J.-P. M., Boensch, D. M., Smith, C. E., Tweet, J. S., & Santucci, V. L. (2023b). *Natchez Trace Parkway and Natchez Trace National Scenic Trail: Paleontological Resource Inventory*. Fort Collins, CO: National Park Service. Natural Resource Report NPS/NATR/NRR—2023/2521. https://doi.org/10.36967/2299391.

Hodnett, J.-P. M., Eshelman, R. E., Gardner, N., & Santucci, V. L. (2023a). *Geology, Pleistocene Paleontology, and Research History of the Cumberland Bone Cave*. Fort Collins, CO: National Park Service. Natural Resource Report NPS/POHE/NRR—2023/2493. https://doi.org/10.36967/2296839.

Hoganson, J. W., & Campbell, J. (1997). Paleontology of Theodore Roosevelt National Park. *North Dakota Geological Survey Newsletter*, **24**(1), 12–23.

Hoganson, J. W., & Murphy, E. C. (2003). *Geology of the Lewis & Clark Trail in North Dakota*. Missoula, MT: Mountain Press Publishing Company.

Hollick, A. (1930). The Upper Cretaceous floras of Alaska. *U.S. Geological Survey Professional Paper*, **159**, 1–123. https://doi.org/10.3133/pp159.

Hollick, A. (1936). The Tertiary floras of Alaska. *U.S. Geological Survey Professional Paper*, **182**, 1–185. https://doi.org/10.3133/pp182.

Holmes, W. H. (1879). Fossil forests of the volcanic tertiary formations of Yellowstone National Park. *U.S. Geological Survey Territories Bulletin*, **2**, 125–132.

Hoots, H. W. (1931). Geology of the eastern part of the Santa Monica Mountains, Los Angeles county, California. *U.S. Geological Survey Professional Paper*, **165-C**, 83–134. https://doi.org/10.3133/pp165C.

Hopkins, D. M., Matthews Jr., J. V., Wolfe, J. A., & Silberman, M. L. (1971). A Pliocene flora and insect fauna from the Bering Strait Region. *Palaeogeography, Palaeoclimatology, Palaeoecology*, **9**(3), 211–231. https://doi.org/10.1016/0031-0182(71)90032-0.

Howard, K. A., Hook, S. J., Phelps, G. A., & Block, D. L. (2003). Geologic map of the Hiller Mountains quadrangle, Clark County, Nevada, and Mohave County, Arizona. *Nevada Bureau of Mines and Geology Map*, **137**, 1 sheet, scale 1:24,000.

Howard, K. A., House, P. K., Dorsey, R. J., & Pearthree, P. A. (2015). River-evolution and tectonic implications of a major Pliocene aggradation on the lower Colorado River: The Bullhead Alluvium. *Geosphere*, **11**, 1–30. https://doi.org/10.1130/GES01059.1.

Hunt, A. P., & Santucci, V. L. (2001). Paleontological resources of Lake Meredith National Recreation Area and Alibates Flint Quarries National Monument, West Texas. *New Mexico Geological Society Guidebook*, **52**, 257–264.

Hunt, R. K., Santucci, V. L., & Kenworthy, J. P. (2007). *Paleontological Resource Inventory and Monitoring: Pacific Islands Network*. Denver, CO: National Park Service. TIC# D-24.

Hunt, R. M. (1990). Taphonomy and sedimentology of Arikaree (lower Miocene) fluvial, eolian, and lacustrine paleoenvironments, Nebraska and Wyoming; a paleobiota entombed in fine-grained volcaniclastic rocks. *Geological Society of America Special Paper*, **244**, 69–111. https://doi.org/10.1130/SPE244-p69.

Izett, G. A., & Obradovich, J. D. (2001). $^{40}Ar/^{39}Ar$ ages of Miocene tuffs in basin-fill deposits (Santa Fe Group, New Mexico, and Troublesome Formation, Colorado) of the Rio Grande Rift System. *The Mountain Geologist*, **38**(2), 77–86.

Jackson, S. T., & Overpeck, J. T. (2000). Responses of plant populations and communities to environmental changes of the late Quaternary. *Paleobiology*, **26**(S4), 194–220. https://doi.org/10.1017/S0094837300026932.

Jacob, A. F. (1973). Depositional environments of Paleocene Tongue River Formation, western North Dakota. *The American Association of Petroleum*

Geologists Bulletin, **57**(6), 1038–1052. https://doi.org/10.1306/83D90E4E-16C7-11D7-8645000102C1865D.

Jacobs, K., & Whitlock, C. (2006). 2000 years of fire history from three sites in Grand Teton National Park. *American Quaternary Association Conference Program and Abstracts*, **19**, 97–98.

Jacobs, K., & Whitlock, C. (2008). A 2000-year environmental history of Jackson Hole, Wyoming, inferred from lake-sediment records. *Western North American Naturalist*, **68**(3), 350–364. https://doi.org/10.3398/1527-0904(2008)68[350:AYEHOJ]2.0.CO;2.

Jain, R. K., & Hall, J. W. (1969). A contribution to the early Tertiary fossil record of the Salviniaceae. *American Journal of Botany*, **56**(5), 527–539. https://doi.org/10.1002/j.1537-2197.1969.tb07566.x.

Jaramillo, C. A. (2018). Evolution of the Isthmus of Panama: Biological, paleoceanographic and paleoclimatological implications. In Hoorn, C., Perrigo, A., & Antonelli, A., eds., *Mountains, Climate and Biodiversity*. Hoboken, NJ: John Wiley & Sons, pp. 323–338.

Johnson, D. L. (1977). The late Quaternary climate of coastal California: Evidence for an ice age refugium. *Quaternary Research*, **8**(2), 154–179. https://doi.org/10.1016/0033-5894(77)90043-6.

Johnson, E., Andeskie, S., Tweet, J. S., & Santucci, V. L. (2023). *Mojave National Preserve: Paleontological Resource Inventory (Public Version)*. Fort Collins, CO: National Park Service. Natural Resource Report NPS/MOJA/NRR—2023/2541. https://doi.org/10.36967/2299742.

Johnson, K. R., & Hickey, L. J. (1990). Megafloral change across the Cretaceous/Tertiary boundary in the northern Great Plains and Rocky Mountains, U.S.A. *Geological Society of America Special Papers*, **247**, 433–444. https://doi.org/10.1130/SPE247-p433.

Johnson, K. R., Nichols, D. J., Attrep, M., & Orth, C. J. (1989). High-resolution leaf-fossil record spanning the Cretaceous/Tertiary boundary. *Nature*, **340** (6236), 708–711. https://doi.org/10.1038/340708a0.

Johnson, R. G. (1962). Mode of marine formation of fossil assemblages of the Pleistocene Millerton Formation of California. *Geological Society of America Bulletin*, **73**(1), 113–130. https://doi.org/10.1130/0016-7606(1962)73[113:MOFOMF]2.0.CO;2.

Keller, A. S., & Reiser, H. N. (1959). Geology of the Mount Katmai area, Alaska. *U.S. Geological Survey Bulletin*, **1058-G**, 261–298. https://doi.org/10.3133/b1058G.

KellerLynn, K. (2007). *Theodore Roosevelt National Park Geologic Resource Evaluation Report*. Denver, CO: National Park Service. Natural Resource Report NPS/NRPC/GRD/NRR—2007/006.

KellerLynn, K. (2014). *Lava Beds National Monument: Geologic Resources Inventory Report*. Fort Collins, CO: National Park Service. Natural Resource Report NPS/NRSS/GRD/NRR—2014/804.

KellerLynn, K. (2016). *Aztec Ruins National Monument: Geologic Resources Inventory Report*. Fort Collins, CO: National Park Service. Natural Resource Report NPS/NRSS/GRD/NRR—2016/1245.

KellerLynn, K. (2019). *Montezuma Castle National Monument: Geologic Resources Inventory Report*. Fort Collins, CO: National Park Service. Natural Resource Report NPS/NRSS/GRD/NRR—2019/2022.

Kenworthy, J. P., & Santucci, V. L. (2003). *Paleontological Resource Inventory and Monitoring: Southwest Alaska Network*. Denver, CO: National Park Service.

Kenworthy, J. P., & Santucci, V. L. (2004). *Paleontological Resource Inventory and Monitoring: National Capital Region*. Denver, CO: National Park Service. TIC# D-289.

Kenworthy, J. P., Santucci, V. L., McNerney, M., & Snell, K. (2005). *Paleontological Resource Inventory and Monitoring: Upper Columbia Basin Network*. Denver, CO: National Park Service. TIC# D-259.

Kenworthy, J. P., Santucci, V. L., & Visaggi, C. C. (2007). *Paleontological Resource Inventory and Monitoring: Gulf Coast Network*. Denver, CO: National Park Service. TIC#D-750.

Kirkham, R. D., & Johnson, M. M. (1929). The Latah Formation in Idaho. *The Journal of Geology*, **37**(5), 483–504. https://doi.org/10.1086/623639.

Knoll, A. H. (1976). The paleontology of the proposed Yukon–Charley Rivers National areas. In Young, S. B., ed., *The Environment of the Yukon–Charley Rivers Area, Alaska*. Wolcott, VT: Center for North Studies, Contributions from The Center for Northern Studies, **9**, pp. 146–172.

Knowlton, F. H. (1899). Fossil flora of the Yellowstone National Park. *U.S. Geological Survey Monograph*, **32**, 651–791. https://doi.org/10.3133/m32.

Knowlton, F. H. (1902). Fossil flora of the John Day Basin, Oregon. *U.S. Geological Survey Bulletin*, **204**, 1–153. https://doi.org/10.3133/b204.

Knowlton, F. H. (1904). Fossil plants from Kukak Bay. In Emerson, B. K., Palache, C., Dall, W. H., Ulrich, E. O., and Knowlton, F. H., eds., *Alaska. Volume IV: Geology and paleontology*. New York: Doubleday, Page & Company, pp. 149–162.

Knowlton, F. H. (1916). A review of the fossil plants in the United States National Museum from the Florissant Lake Beds at Florissant, Colorado: With descriptions of new species and list of type-specimens. *Proceedings of the United States National Museum*, **51**(2151), 241–297. https://doi.org/10.5479/si.00963801.51-2151.241.

Knowlton, F. H. (1926). Flora of the Latah Formation of Spokane, Washington, and Coeur d'Alene, Idaho. *U.S. Geological Survey Professional Paper*, **140**, 17–81. https://doi.org/10.3133/pp140.

Knox, A. S. (1966). The Walker Interglacial Swamp, Washington, D.C. *Journal of the Washington Academy of Sciences*, **56**(1), 1–8.

Korasidis, V. A., & Wing, S. L. (2023). Palynofloral change through the Paleocene–Eocene Thermal Maximum in the Bighorn Basin, Wyoming. *Paleoceanography and Paleoclimatology*, **38**, e2023PA004741, 1–25. https://doi.org/10.1029/2023PA004741.

Kottkamp, S., Santucci, V. L., Tweet, J. S., De Smet, J., & Starck, E. (2020). *Agate Fossil Beds National Monument: Paleontological Resources Management Plan (Public Version)*. Fort Collins, CO: National Park Service. Natural Resource Report NPS/AGFO/NRR—2020/2172. https://doi.org/10.36967/nrr-2278685.

Kümmel, H. B., & Poland, H. M. (1910). Records of wells in New Jersey, 1905–1909. In *Annual Report of the State Geologist for the Year 1909*. Trenton, NJ: Geological Survey of New Jersey, pp. 69–100.

Kuntz, M. A., Lefevre, R. H., Champion, D. E., & Skipp, B. A. (1989). Geologic map of the Inferno Cone quadrangle, Butte County, Idaho. *U.S. Geological Survey Geologic Quadrangle Map*, **1632**, 1 sheet, scale 1:24,000. https://doi.org/10.3133/gq1632.

LaGarry, H. E., Terry, D. O., & Hunt, R. M. (1998). Lithostratigraphic revision and redescription of the Brule Formation (White River Group) of northwestern Nebraska. *Geological Society of America Special Paper*, **325**, 63–92. https://doi.org/10.1130/0-8137-2325-6.63.

Lamb, M., Beard, L. S., Dragos, M., et al. (2018). Provenance and paleogeography of the 25-17 Ma Rainbow Gardens Formation; evidence for tectonic activity at ca. 19 Ma and internal drainage rather than throughgoing paleorivers on the southwestern Colorado Plateau. *Geosphere*, **14**(4), 1592–1617. https://doi.org/10.1130/GES01127.1.

Lander, E. B., compiler. (2011). *Stratigraphy, Biostratigraphy, Biochronology, Geochronology, Magnetostratigraphy, and Plate Tectonic History of the Early Middle Eocene to Late Early Miocene Sespe, Vaqueros, and Lower Topanga Formations, East-Central Santa Monica Mountains, Los Angeles County, Southern California*. Los Angeles: Natural History Museum of Los Angeles County. Western Association of Vertebrate Paleontologists 2011 annual meeting field trip volume and guidebook.

Lander, E. B., Hemming, S. R., & Nyborg, T. (2022). Titus Canyon Formation, Death Valley National Park, California and Nevada—tuffaceous marker beds, $^{40}Ar/^{39}Ar$ dates, and constraints on age of Titus Canyon Fauna. In

Miller, D. M., ed., *Volcanoes in the Mojave: 2022 Desert Symposium Field Guide and Proceedings*. Zzyzx, CA: Desert Studies Center, pp. 240–241.

Lange, A. (1957). Studies on the origin of Montezuma Well and cave, Arizona. *Cave Studies*, **9**, 39–53.

Lanik, A., Swanson, D. K., & Karpilo, R. D. (2019). *Bering Land Bridge National Preserve: Geologic Resources Inventory Report*. Fort Collins, CO: National Park Service. Natural Resource Report NPS/NRSS/GRD/NRR—2019/2024.

Laughlin, A. W., Charles, R. W., Reid, K., & White, C. (1993). Field-trip guide to the geochronology of El Malpais National Monument and the Zuni-Bandera volcanic field, New Mexico. *New Mexico Bureau of Geology and Mineral Resources Bulletin*, **149**, 1–23.

Laughlin, A. W., & Perry, F. V. (1997). Photographic atlas of volcanic features. *New Mexico Bureau of Geology and Mineral Resources Bulletin*, **156**, 13–24.

Laughlin, A. W., Poths, J., Healey, H. A., Reneau, S., & Woldegabriel, G. (1994). Dating of Quaternary basalts using the cosmogenic ^{3}He and ^{14}C methods with implications for excess ^{40}Ar. *Geology*, **22**(2), 135–138. https://doi.org/10.1130/0091-7613(1994)022%3C0135:DOQBUT%3E2.3.CO;2.

Lawson, D. A. (1972). *Paleoecology of the Tornillo Formation, Big Bend National Park, Brewster County, Texas*. Austin: University of Texas, master's thesis.

Lawson, D. A., & Wiles, G. (2009). *Paleoclimate of the Last 10,000 Years, Glacier Bay National Park and Preserve: Progress Understanding Climate Change in Southeast Alaska*. Gustavus, AK: Glacier Bay National Park and Preserve. 2009 Annual Report.

Lawson, D. A., Wiles, G., & Finnegan, D. (2007). A dendroclimatic record of paleoclimate od the last 10,000 years, Glacier Bay National Park and Preserve: Progress understanding climate change in southeast Alaska. *U.S. Army Research*, **22**, 1–29.

Lehman, T. M., Wick, S. L., Beatty, H. L., Straight, W. H., & Wagner, J. R. (2018). Stratigraphy and depositional history of the Tornillo Group (Upper Cretaceous–Eocene) of West Texas. *Geosphere*, **14**(5), 2206–2244. https://doi.org/10.1130/ges01641.1.

Leopold, E. B., & Clay-Poole, S. T. (2001). Fossil leaf and pollen floras of Colorado compared: Climatic implications. *Proceedings of the Denver Museum of Nature and Science, Series 4*, **1**, 17–69.

Leopold, E. B., & Denton, M. F. (1987). Comparative age of grassland and steppe east and west of the northern rocky mountain. *Annals of the Missouri Botanical Garden*, **74**(4), 841–867. https://doi.org/10.2307/2399452.

Leopold, E. B., & Liu, G. (1994). A long pollen sequence of Neogene age, Alaska Range. *Quaternary International*, **22/23**, 103–140. https://doi.org/10.1016/1040-6182(94)90009-4.

Leopold, E. B., Liu, G., Love, J. D., & Love, D. W. (2007). Plio-Pleistocene climatic transition and the lifting of the Teton Range, Wyoming. *Quaternary Research*, **67**(1), 1–11. https://doi.org/10.1016/j.yqres.2006.10.006.

Leopold, E. B., & MacGinitie, H. D. (1972). Development and affinities of Tertiary floras in the Rocky Mountains. In Graham, A. ed., *Floristics and Paleofloristics of Asia and Eastern North America*. Amsterdam: Elsevier Publishing Co., pp. 147–189.

Leopold, E. B., & Meyer, H. W. (2012). *Saved in Time: The Fight to Establish Florissant Fossil Beds National Monument, Colorado*. Albuquerque: University of New Mexico Press.

Lesquereux, L. (1872). Fossil flora. In F. V. Hayden, ed., *Preliminary Report of the United States Geological Survey of Montana and Portions of Adjacent Territories, Being a Fifth Annual Report of Progress*. Washington, DC: Department of the Interior, pp. 283–318. https://doi.org/10.3133/70038938.

Lesquereux, L. (1878). Contributions to the fossil flora of the Western Territories. Part II. The Tertiary flora. *Report of the United States Geological Survey of the Territories*, **7**, 1–366. https://doi.org/10.3133/70038957.

Lesquereux, L. (1883). Contributions to the fossil flora of the Western Territories. Part III. The Cretaceous and Tertiary floras. *Report of the United States Geological and Geographical Survey of the Territories*, **8**, 1–280. https://doi.org/10.3133/70038958.

Logan, G. A., Boon, J. J., & Eglinton, G. (1993). Structural biopolymer preservation in Miocene leaf fossils from the Clarkia site, northern Idaho. *Proceedings of the National Academy of Sciences*, **90**(6), 2246–2250. https://doi.org/10.1073/pnas.90.6.2246.

Long, C. J., Power, M. J., & McDonald, B. (2011). Millennial-scale fire and vegetation history from a mesic hardwood forest of southeastern Wisconsin, USA. *Journal of Quaternary Science*, **26**(3), 318–325. https://doi.org/10.1002/jqs.1456.

Longwell, C. R. (1936). Geology of the Boulder Reservoir floor, Arizona-Nevada. *Geological Society of America Bulletin*, **47**(9), 1393–1476. https://doi.org/10.1130/GSAB-47-1393.

Loope, D. B., Mason, J. A., Bao, H., Kettler, R. M., & Zanner, C. W. (2005). Deformation structures and an alteration zone linked to deposition of volcanogenic sulphate in an ancient playa (Oligocene of Nebraska, USA).

Sedimentology, **52**(1), 123–139. https://doi.org/10.1111/j.1365-3091.2004.00689.x.

Loope, W. L., Fisher, T. G., Jol, H. M., Anderton, J. B., & Blewett, W. L. (2004). A Holocene history of dune-mediated landscape change along the southeastern shore of Lake Superior. *Geomorphology*, **61**(3–4), 303–322. https://doi.org/10.1016/j.geomorph.2004.01.005.

Lowe, A. (2024). *Plant Community and Climatic Response to Miocene Environmental Change in the Pacific Northwest (USA): Paleoecological Tools and Applications*. Seattle: University of Washington, doctoral dissertation.

Lüthi, D., Le Floch, M., Bereiter, B., et al. (2008). High-resolution carbon dioxide concentration record 650,000–800,000 years before present. Nature, **453**, 379–382. https://doi.org/10.1038/nature06949.

Lyson, T. R., Miller, I. M., Bercovici, A. D., et al. (2019). Exceptional continental record of biotic recovery after the Cretaceous–Paleogene mass extinction. *Science*, **366**(6468), 977–983. https://doi.org/10.1126/science.aay2268.

Mabery, K., compiler. (1997). Natural history of El Malpais National Monument. *New Mexico Bureau of Geology and Mineral Resources Bulletin*, **156**, 1–185. https://doi.org/10.58799/B-156.

MacGinitie, H. D. (1953). Fossil plants of the Florissant beds, Colorado. *Carnegie Institution of Washington Publication*, **599**, 1–198.

MacGinitie, H. D. (1962). The Kilgore flora: A Late Miocene flora from northern Nebraska. *University of California Publications in Geological Sciences*, **35**(1), 67–158.

MacGinitie, H. D. (1969). The Eocene Green River flora of northwestern Colorado and northeastern Utah. *University of California Publications in Geological Sciences*, **83**, 1–203.

MacGinitie, H. D. (1974). An early Middle Eocene flora from the Yellowstone-Absaroka Volcanic Province, northwestern Wind River Basin, Wyoming. *University of California Publications in Geological Sciences*, **108**, 1–103.

MacNeil, F. S., Wolfe, J. A., Miller, D. J., & Hopkins, D. M. (1961). Correlation of Tertiary formations of Alaska. *American Association of Petroleum Geologists Bulletin*, **45**(11), 1801–1809.

Madsen, D. B., Sarna-Wojcicki, A. M., & Thompson, R. S. (2002). A late Pleistocene tephra layer in the southern Great Basin and Colorado Plateau derived from Mono Craters, California. *Quaternary Research*, **57**(3), 382–390. https://doi.org/10.1006/qres.2002.2326.

Madsen, S. K. (2010). *Exploring Desert Stone: John N. Macomb's 1859 Expedition to the Canyonlands of the Colorado*. Salt Lake City: Utah State University Press.

Magoon, L. B., Adkison, W. L., Chmelik, F. B., et al. (1976a). Hydrocarbon potential, geologic hazards and infrastructure for exploration and development of the lower Cook Inlet, Alaska. *U.S. Geological Survey Open-File Report*, **76-449**, 1–124. https://doi.org/10.3133/ofr76449.

Magoon, L. B., Adkison, W. L., & Egbert, R. M. (1976b). Map showing geology, wildcat wells, Tertiary plant fossil localities, K-Ar age dates, and petroleum operations, Cook Inlet area, Alaska. *U.S. Geological Survey Miscellaneous Investigations Series Map*, **1019**, 3 sheets, scale 1:250,000. https://doi.org/10.3133/i1019.

Manchester, S. R. (1981). Fossil plants of the Eocene Clarno Nut Beds. *Oregon Geology*, **43**, 75–81.

Manchester, S. R. (1987). Fossil history of the Juglandaceae. *Missouri Botanical Garden Monograph*, **21**, 1–137.

Manchester, S. R. (1989a). Attached reproductive and vegetative remains of the extinct American-European genus *Cedrelospermum* (Ulmaceae) from the early Tertiary of Utah and Colorado. *American Journal of Botany* **76**, 256–276. https://doi.org/10.1002/j.1537-2197.1989.tb11309.x.

Manchester, S.R. (1989b). Early history of the Juglandaceae. *Plant Systematics and Evolution*, **162**, 231–250.

Manchester, S. R. (1991). *Cruciptera*, a new juglandaceous winged fruit from the Eocene and Oligocene of western North America. *Systematic Botany*, **16**(4), 715–725. https://doi.org/10.2307/2418873.

Manchester, S. R. (1992). Flowers, fruits, & pollen of *Florissantia*, an extinct malvalean genus from the Eocene and Oligocene of western North America. *American Journal of Botany*, **79**, 996–1008.

Manchester, S. R. (1994). Fruits and seeds of the Middle Eocene Nut Beds flora, Clarno Formation, Oregon. *Palaeontographica Americana*, **58**, 1–205.

Manchester, S. R. (1999). Biogeographical relationships of North American Tertiary floras. *Annals of the Missouri Botanical Garden*, **86**(2), 472–522. https://doi.org/10.2307/2666183.

Manchester, S. R. (2001). Update on the megafossil flora of Florissant, Colorado. *Proceedings of the Denver Museum of Nature and Science, Series 4*, **1**, 137–161. https://doi.org/10.1002/j.1537-2197.1992.tb13689.x.

Manchester, S. R. (2014). Revisions to Roland Brown's North American Paleocene flora. *Acta Musei Nationalis Pragae Series B, Historia Naturalis*, **70**(3–4), 153–210. https://doi.org/10.14446/AMNP.2014.153.

Manchester, S. R., & Crane, P. R. (1983). Attached leaves, inflorescences, and fruits of *Fagopsis*, an extinct genus of fagaceous affinity from the Oligocene Florissant fauna of Colorado, USA. *American Journal of Botany*, **70**, 1147–1164. https://doi.org/10.1002/j.1537-2197.1983.tb12464.x.

Manchester, S. R., & Crane, P. R. (1987). A new genus of Betulaceae from the Oligocene of western North America. *Botanical Gazette*, **148**, 263–273. https://doi.org/10.1086/337654.

Manchester, S. R., Crane, P. R., & Dilcher, D. L. (1991). *Nordenskioldia* and *Trochodendron* (Trochodendraceae) from the Miocene of northwestern North America. *Botanical Gazette*, **152**, 357–368. https://doi.org/10.1086/337898.

Manchester, S. R., Dilcher, D. L., & Wing, S. L. (1998). Attached leaves and fruits of myrtaceous affinity from the middle Eocene of Colorado, USA. *Review of Palaeobotany and Palynology*, **102**, 153–163.

Manchester, S. R., & Donoghue, M. J. (1995). Winged fruits of Linnaeeae (Caprifoliaceae) in the Tertiary of western North America: *Diplodipelta* gen. nov. *International Journal of Plant Sciences*, 156(5), 709–722. https://doi.org/10.1086/297293.

Manchester, S. R., Judd, W. S., & Kodrul, T. (2024). First recognition of the extinct eudicot genus *Palibinia* in North America: Leaves and fruits of *Palibinia comptonifolia* (R.W.Br.) comb. nov. from the Eocene of Utah and Colorado, USA. *Journal of Systematics and Evolution*, 62(1), 149–163. https://doi.org/10.1111/jse.13011.

Manchester, S. R., Pigg, K. B., & Crane, P. R. (2004). *Palaeocarpinus dakotensis* sp. n. (Betulacceae: Coryloideae) and associated staminate catkins, pollen and leaves from the Paleocene of North Dakota. *International Journal of Plant Science*, **165**, 1135–1148. https://doi.org/10.1086/423870.

Marincovich, L. Jr. (1988). Late Middle Eocene mollusks of the Tolstoi Formation, Alaska Peninsula, Alaska, and correlations with faunas from California to the Far-eastern U.S.S.R. In Filewicz, M. V., & Squires, R. L., eds., *Paleogene Stratigraphy, West Coast of North America*. Los Angeles: Society of Economic Paleontologists and Mineralogists, Pacific Section, pp. 265–281.

Marincovich, L., Jr., & McCoy Jr., S. (1984). An overview of Paleogene molluscan biostratigraphy and paleoecology of the Gulf of Alaska Region. *Palaeogeography, Palaeoclimatology, Palaeoecology*, **47**, 91–102. https://doi.org/10.1016/0031-0182(84)90082-8.

Martin, J. E. (1984). *Vertebrate Remains from Wind Cave, Custer County, South Dakota*. Internal report to Wind Cave National Park, DS.

Martin, J. E. (1988). *Geology and Vertebrate Paleontology of the Klukas Localities*. Internal report to Wind Cave National Park, SD.

Martin, L. D., & Bennett, D. K. (1977). The burrows of the Miocene beaver *Palaeocastor*, western Nebraska, U.S.A. *Palaeogeography, Palaeoclimatology, Palaeoecology*, **22**, 173–193. https://doi.org/10.1016/0031-0182(77)90027-X.

Mason, H. L. (1934). Pleistocene flora of the Tomales Formation. *Carnegie Institution of Washington Publication*, **415**, 81–179.

Mason, O. K., & Ludwig, S. L. (1990). Resurrecting beach ridge archeology, parallel depositional records from St. Lawrence Island and Cape Krusenstern, western Alaska. *Geoarcheology*, **5**, 349–373. https://doi.org/10.1002/gea.3340050404.

Maxwell, R. A., Lonsdale, J. T., Hazzard, R. T., & Wilson, J. A. (1967). *Geology of Big Bend National Park, Brewster County, Texas*. Austin: University of Texas Bureau of Economic Geology.

McCartan, L., Tiffney, B. H., Wolfe, J. A., et al. (1990). Late Tertiary floral assemblage from upland gravel deposits of the southern Maryland coastal plain. *Geology*, **18**, 311–314. https://doi.org/10.1130/0091-7613(1990)018%3C0311:LTFAFU%3E2.3.CO;2.

McElwain, J. C., & Punyasena, S. W. (2007). Mass extinction events and the plant fossil record. *Trends in Ecology & Evolution*, **22**(10), 548–557.

McInerney, F. A., & Wing, S. L. (2011). The Paleocene–Eocene Thermal Maximum: A perturbation of carbon cycle, climate, and biosphere with implications for the future. *Annual Review of Earth and Planetary Sciences*, **39**, 489–516. https://doi.org/10.1146/annurev-earth-040610-133431.

McIver, E. E., & Basinger, J. F. (1993). Fossil flora of the Ravenscrag Formation (Paleocene), southwestern Saskatchewan, Canada. *Palaeontographica Canadiana*, **10**, 1–167.

McKee, T. M. (1970). Preliminary report on fossil fruits and seeds from the mammal quarry of the Clarno formation, Oregon. *The Ore Bin*, **32**, 117–132.

McKenna, M. C. (1975). Fossil mammals and Early Eocene North Atlantic land continuity. *Annals of the Missouri Botanical Garden*, **62**(2), 335–353. https://doi.org/10.2307/2395200.

McKenna, M. C. (1983). Cenozoic paleogeography of North Atlantic land-bridges. In Bott, M. H P., Saxov, S., Talwani, M., & Thiede, J., eds., *Structure and Development of the Greenland-Scotland Ridge*. New York: Plenum, pp. 351–399.

McLeroy, C. A., & Anderson, R. Y. (1966). Laminations of the Oligocene Florissant Lake deposits, Colorado. *Geological Society of America Bulletin*, **77**, 605–618. https://doi.org/10.1130/0016-7606(1966)77[605:LOTOFL]2.0.CO;2.

McMurran, D. M., & Manchester, S. R. (2010). *Lagokarpos lacustris*, a new winged fruit from the Paleogene of western North America. *International Journal of Plant Sciences*, **171**(2), 227–234. https://doi.org/10.1086/648994.

Meehan, T. J. (1994). Sediment analysis of the middle Whitney Member: Climatic implications for the upper Oligocene of western Nebraska. *TER-QUA Symposium Series: Institute for Tertiary-Quaternary Studies*, **2**, 57–87.

Mercer, H. C. (1899). The bone cave at Port Kennedy, Pennsylvania, and its partial excavation in 1894, 1895, and 1896. *Journal of the Academy of Natural Sciences of Philadelphia*, **11**, 269–286.

Mertie, J. B., Jr. (1930). Geology of the Eagle–Circle district, Alaska. *U.S. Geological Survey Bulletin*, **816**, 1–168. https://doi.org/10.3133/b816.

Mertie, J. B., Jr. (1942). Tertiary deposits of the Eagle–Circle district, Alaska. *U.S. Geological Survey Bulletin*, **917-D**, 213–264. https://doi.org/10.3133/b917D.

Meyer, H. W. (1986). *An Evaluation of the Methods for Estimating Paleoaltitudes Using Tertiary Floras from the Rio Grande Rift Vicinity, New Mexico and Colorado*. Berkeley: University of California, doctoral dissertation.

Meyer, H. W. (2003). *The Fossils of Florissant*. Washington: Smithsonian Books.

Meyer, H. W., & Manchester, S. R. (1997). The Oligocene Bridge Creek flora of the John Day Formation, Oregon. *University of California Publications in Geological Science*, **141**, 1–195.

Meyer, H. W., Wasson, M. S., & Frakes, B. J. (2008). Development of an integrated paleontological database and Web site of Florissant collections, taxonomy, and publications. *Geological Society of America Special Paper*, **435**, 159–177. https://doi.org/10.1130/2008.2435(11).

Millar, C. I. (1998). Reconsidering the conservation of Monterey pine. *Fremontia* **26**(3), 13–17.

Miller, K. G., Sugarman, P. J., Browning, J. V., et al. (1998). Bass River Site. *Proceedings of the Ocean Drilling Program, Initial Reports*, **174AX**, 5–43.

Miller, K. G., Sugarman, P. J., Browning, J. V., et al. (1999). Ancora site. *Proceedings of the Ocean Drilling Program, Initial Reports*, **174AX** (supplement), 1–65.

Miller, K. G., Sugarman, P., van Fossen, M., et al. (1994). Island Beach site report. *Proceedings of the Ocean Drilling Program, Initial Reports*, **150X**, 5–33.

Minard, J. P. (1969). Geology of the Sandy Hook Quadrangle in Monmouth County, New Jersey. *U.S. Geological Survey Bulletin*, **1276**, 1–43. https://doi.org/10.3133/b1276.

Mixon, R. B., Berquist, C. R., Newell, W. L., & Johnson, G. H. (1989). Geologic map and generalized cross sections of the coastal plain and adjacent parts of the Piedmont, Virginia. *U.S. Geological Survey Miscellaneous Investigations Series Map*, **2033**, 1 map, scale 1:250,000. https://doi.org/10.3133/i2033.

Miyaoka, R. T. (1990). Fossil locality map and fossil data for the southeastern Charley River Quadrangle, east-central Alaska. *U.S. Geologic Survey Miscellaneous Field Studies Map*, **2007**, 1 sheet, scale 1:100,000. https://doi.org/10.3133/mf2007.

Montes, C., Cardona, A., Jaramillo, C., et al. (2015). Middle Miocene closure of the Central American Seaway. *Science*, **348**(6231), 226–229. https://doi.org/10.1126/science.aaa2815.

Morgan, A. V., & Morgan, A. (1979). The fossil Coleoptera of the Two Creeks forest bed, Wisconsin. *Quaternary Research*, **12**(2), 226–240. https://doi.org/10.1016/0033-5894(79)90059-0.

Muir, J. (1915). *Travels in Alaska*. Boston and New York: Houghton Mifflin Co.

Mulvey, W. E., Currey, D. R., & Lindsay, L. W. (1984). Southernmost occurrence of late Pleistocene glaciation in Utah: Brian Head-Sidney Peaks area, Markagaunt Plateau. *Encyclia*, **61**, 97–103.

Mustoe, G. E. (2019). Lower Eocene footprints from northwest Washington, USA. Part 1: Reptile tracks. *Geosciences*, **9**(7), 321, 1–19. https://doi.org/10.3390/geosciences9070321.

Mustoe, G. E., Dillhoff, R. M., & Dillhoff, T. A. (2007). Geology and paleontology of the early Tertiary Chuckanut Formation. *Geological Society of America Field Guide*, **9**, 121–135. https://doi.org/10.1130/2007.fld009(06).

Mustoe, G. E., & Gannaway, W. L. (1995). Palm fossils from northwest. *Washington Geology*, **23**(2), 21–26.

Mustoe, G. E., & Gannaway, W. L. (1997). Paleogeography and paleontology of the early Tertiary Chuckanut Formation, northwest Washington. *Washington Geology*, **25**, 1–18.

Nash, B. P., & Perkins, M. E. (2012). Neogene fallout tuffs from the Yellowstone Hotspot in the Columbia Plateau Region, Oregon, Washington, and Idaho, USA. *PLoS ONE*, **7**(10): e44205. https://doi.org/10.1371/journal.pone.0044205.

National Park Service (NPS). (2015). Unpublished digital geologic map of Bering Land Bridge NP and vicinity, Alaska. Denver, CO: National Park Service. https://irma.nps.gov/DataStore/Reference/Profile/2222911.

Nelson, M., Antonioni, M., Santucci, V. L., & Tweet, J. S. (2019). *National Capital Parks-East Paleontological Resources Inventory*. Fort Collins, CO: National Park Service. Natural Resource Report NPS/NACE/NRR—2019/1862.

Newberry, J. S. (1898). The later extinct floras of North America. *U.S. Geological Survey Monograph*, **35**, 1–295. https://doi.org/10.3133/m35.

Nichols, D. J., & Johnson, K. R. (2008). *Plants and the K–T Boundary*. Cambridge: Cambridge University Press.

Niklas, K. J., & Brown, R. M., Jr. (1981). Ultrastructural and paleobiochemical correlations among fossil leaf tissues from the St. Maries River (Clarkia) area, northern Idaho, USA. *American Journal of Botany*, **68**(3), 332–341.

Nixon, D. A., & LaGarry-Guyon, H. E. (1993). New trackway site in the White River Group type section at Toadstool Park. Nebraska: Paleoecology of an Oligocene braided stream, riparian woodland, and adjacent grassland. *Journal of Vertebrate Paleontology*, **13**(supplement to 3), 50A.

Nyborg, T. G., & Santucci, V. L. (1999). *Death Valley National Park Paleontological Survey*. Lakewood, CO: National Park Service.

O'Dea, A., Lessios, H. A., Coates, A. G., et al. (2016). Formation of the Isthmus of Panama. *Science Advances*, **2**(8), e1600883. https://doi.org/10.1126/sciadv.1600883.

Oldale, R. N. (1968). Geologic map of the Wellfleet quadrangle, Barnstable County, Cape Cod, Massachusetts. *U.S. Geological Survey Geologic Quadrangle Map*, **911**, 1 map, scale 1:24,000. https://doi.org/10.3133/gq911.

Oldale, R. N. (2001). *Cape Cod, Martha's Vineyard, and Nantucket: The Geologic Story*. Yarmouth Port, MA: On Cape Publications.

Otto, A., Simoneit, B. R. T., & Rember, W. C. (2003). Resin compounds from the seed cones of three fossil conifer species from the Miocene Clarkia flora, Emerald Creek, Idaho, USA, and from related extant species. *Review of Palaeobotany and Palynology*, **126**(3–4), 225–241. https://doi.org/10.1016/S0034-6667(03)00088-5.

Owen, D. E., & Melander, S. M. (2014). *Geology of Craters of the Moon*. Arco, ID: Craters of the Moon National History Association. https://irma.nps.gov/DataStore/Reference/Profile/2252227.

Owens, J. P., Bybell, L. M., Paulachok, G., et al. (1988). Stratigraphy of the Tertiary sediments in a 945-foot-deep corehole near Mays Landing in the southeastern New Jersey Coastal Plain. *U.S. Geological Survey Professional Paper*, **1484**, 1–39. https://doi.org/10.3133/pp1484.

Parker, B. C., & Dawson, E. Y. (1965). Non-calcareous marine algae from California Miocene deposits. *Nova Hedwigia* **10**(1–2), 273–295.

Parker, J. D. (1983). Lower Paleocene to lower Eocene, nonmarine to deep-marine strata of the Simi Hills, Ventura County, California. In Squires, R. L., & Filewicz, M. V., eds., *Cenozoic Geology of the Simi Valley Area*. Los Angeles: Society of Economic Paleontologists and Mineralogists, Pacific Section, 3–22.

Parrish, J. T., Fiorillo, A. R., Jacobs, B. F., Currano, E. D., & Wheeler, E. A. (2010). The Ketavik Formation: A new stratigraphic unit and its implications for the Paleogene paleogeography and paleoclimate of southwestern Alaska. *Palaeogeography, Palaeoclimatology, Palaeoecology*, **295**(3–4), 348–362. https://doi.org/10.1016/j.palaeo.2009.09.008.

Peppe, D. J. (2010). Megafloral change in the early and middle Paleocene in the Williston Basin, North Dakota, USA. *Palaeogeography, Palaeoclimatology, Palaeoecology*, **298**(3–4), 224–234. www.sciencedirect.com/science/article/abs/pii/S0031018210005985.

Petit, J., Jouzel, J., Raynaud, D., et al. (1999). Climate and atmospheric history of the past 420,000 years from the Vostok ice core, Antarctica. *Nature*, **399**, 429–436. https://doi.org/10.1038/20859.

Pierce, H. G., & Constenius, K. N. (2001). Late Eocene-Oligocene nonmarine mollusks of the northern Kishenehn Basin, Montana and British Columbia. *Annals of Carnegie Museum* **70**(1), 1–112.

Pierce, H. G., & Constenius, K. N. (2014). Terrestrial and aquatic mollusks of the Eocene Kishenehn Formation, Middle Fork Flathead River, Montana. *Annals of Carnegie Museum*, **82**(4), 305–332. https://doi.org/10.2992/007.082.0401.

Pierce, K. L., & Good, J. D. (1992). Field guide to the Quaternary geology of Jackson Hole, Wyoming. *U.S. Geological Survey Open-File Report*, **92-504**, 1–54. https://doi.org/10.3133/ofr92504.

Pigati, J. S., Springer, K. B., Honke, J. S., et al. (2023). Independent age estimates resolve the controversy of ancient human footprints at White Sands. *Science*, **382**(6666), 73–75. https://doi.org/10.1126/science.adh5007.

Pigg, K. B., & DeVore, M. L. (2010). Floristic composition and variation in late Paleocene to Early Eocene floras in North America. *Bulletin of Geosciences*, **85**(1), 135–154. https://doi.org/10.3140/bull.geosci.1136.

Pigg, K. B., DeVore, M. L., & Taylor, W. (2023). New features of *Cyclocarya brownii* Manchester & Dilcher from the Late Paleocene of North Dakota, USA. *International Journal of Plant Sciences*, **184**(4), 282–303. https://doi.org/10.1086/724496.

Pigg, K. B., Manchester, S. R., & DeVore, M. L. (2008). Fruits of Icacinaceae (Tribe Iodeae) from the Late Paleocene of western North America. *American Journal of Botany*, **95**(7), 824–832. https://doi.org/10.3732/ajb.2007340.

Pigg, K. B., Manchester, S. R., & Wehr, W. C. (2003). *Corylus*, *Carpinus*, and *Palaeocarpinus* (Betulaceae) from the Middle Eocene Klondike Mountain and Allenby Formations of Northwestern North America. *International Journal of Plant Sciences*, **164**(5), 807–822.

Plafker, G. (1967). Geologic map of the Gulf of Alaska Tertiary Province, Alaska. *U.S. Geological Survey Miscellaneous Investigations Series Map*, **484**, 1 sheet, scale 1:500,000. https://doi.org/10.3133/i484.

Platen, P. L. (1908). Untersuchungen fossiler Hölzer aus dem Westen der Vereinigten Staaten von Nordamerika. *Sitzungsberichte der Naturforschende Gesellschaft zu Leipzig*, **34**, 1–155.

Polenz, M., Slaughter, S. L., Dragovich, J. D., & Thorsen, G. W. (2005a). Geologic map of the Ebey's Landing National Historical Reserve, Island County, Washington. *Washington Division of Geology and Earth Resources Open-File Report*, **2005-2**, 1 sheet, scale 1:24,000.

Polenz, M., Slaughter, S. L., & Thorsen, G. W. (2005b). Geologic map of the Coupeville and part of the Port Townsend North 7.5-minute quadrangles, Island County, Washington. *Washington Division of Geology and Earth Resources Geologic Map*, **GM-58**, 1 sheet, scale 1:24,000.

Poore, R. Z., Pavich, M. J., & Grissino-Mayer, H. D. (2005). Record of the North American Southwest Monsoon from Gulf of Mexico sediment cores. *Geology*, **33**(3), 209–212. https://doi.org/10.1130/G21040.1.

Potzger, J. E. (1945). The pine barrens of New Jersey: A refugium during Pleistocene times. *Butler University Botanical Studies*, **7**, 182–196.

POWO. (2023). *Plants of the World Online*. Facilitated by the Royal Botanic Gardens, Kew. Published on the Internet; https://powo.science.kew.org/.

Pringle, P. T. (2008). Roadside geology of Mount Rainier National Park and vicinity. *Washington Division of Geology and Earth Resources Information Circular*, **107**, 1–190.

Pringle, P. T., & Scott, K. M. (2001). Postglacial influence of volcanism on the landscape and environmental history of the Puget Lowland, Washington: A review of geologic literature and recent discoveries, with emphasis on the landscape disturbances associated with lahars, lahar runouts, and associated flooding. In *Puget Sound Research 2001, Proceedings*. Seattle: Washington State Puget Sound Water Quality Action Team, unpaginated.

Prothero, D. R., & Emry, R. J. (2004). The Chadronian, Orellan, and Whitneyan North American Land Mammal ages. In Woodburne, M. O., ed., *Late Cretaceous and Cenozoic Mammals of North America*. New York: Columbia University Press, pp. 156–168. https://doi.org/10.7312/wood13040-007.

Prothero, D. R., & Whittlesey, K. E. (1995). Redefinition of the Chadronian/Orellan boundary in the White River Group: Implications for Eocene-Oligocene faunal change. *Geological Society of America Abstracts with Programs*, **27**(6), A-320.

Ramírez-Barahona, S., Sauquet, H., & Magallón, S. (2020). The delayed and geographically heterogeneous diversification of flowering plant families.

Nature Ecology & Evolution, **4**(9), 1232–1238. https://doi.org/10.1038/s41559-020-1241-3.

Rawling, J. E., III, and Fredlund, G. G. (1999). Episodic mid-late Holocene eolian activity evidenced by buried soils in cliff-dunes at Sheep Mountain Table, Badlands National Park, South Dakota. *Geological Society of America Abstracts with Programs*, **31**(7), 54.

Rember, W. C. (1991). *Stratigraphy and Paleobotany of Miocene Lake Sediments Near Clarkia, Idaho*. Moscow: University of Idaho, doctoral dissertation.

Rensberger, J. M. (1976). *Paleontologic Information, John Day Fossil Beds National Monument*. Internal report to John Day Fossil Beds National Monument.

Repenning, C. A. (1974). Report on referred fossils [shipment A-74-26 M]. USGS memo to T. D. Hamilton, October 29, 1974.

Retallack, G. J. (1981). Paleopedological perspectives on the functional morphology and taphonomy of vertebrate fossil assemblages. *Geological Society of America Abstracts with Programs*, **13**(7), 537.

Retallack, G. J. (1983). Late Eocene and Oligocene paleosols from Badlands National Park, South Dakota. *Geological Society of America Special Paper*, **193**, 1–82. https://doi.org/10.1130/SPE193.

Retallack, G. J. (1984). Completeness of the rock and fossil record: Some estimates using fossil soils. *Paleobiology*, **10**(1), 59–78. https://doi.org/10.1017/S0094837300008022.

Retallack, G. J. (1992). Paleosols and changes in climate and vegetation across the Eocene-Oligocene boundary. In Prothero, D. R., & Berggren, W. A., eds., *Eocene-Oligocene Climatic and Biotic Evolution*. Princeton, NJ: Princeton University Press, pp. 383–398.

Retallack, G. J. (1997). Neogene expansion of the North American prairie. *Palaios*, **12**(4), 380–390. https://doi.org/10.2307/3515337.

Reynolds, R. E., Hunt, R., & Albright, B. (1995). Rhinoceros in Lanfair Valley. *San Bernardino County Museum Quarterly*, **42**(3), 107–110.

Richards, H. G., & Thurber, D. L. (1966). Pleistocene age determinations from California and Oregon. *Science*, **152**(3725), 1091–1092. https://doi.org/10.1126/science.152.3725.1091.

Richter, D. H., Preller, C. C., Labay, K. A., & Shew, N. B. (2006). Geologic map of the Wrangell–Saint Elias National Park and Preserve, Alaska. *U.S. Geological Survey Scientific Investigations Map*, **2877**, 1–14, 1 sheet, scale 1:250,000. https://doi.org/10.3133/sim2877.

Riedel, J. L. (2007). *Late Pleistocene Glacial and Environmental History of the Skagit Valley, Washington and British Columbia.* Vancouver: Simon Fraser University, doctoral dissertation.

Riedel, J. L., Pringle, P. T., & Schuster, R. L. (2001). Deposition of Mount Mazama tephra in a landslide-damned lake on the upper Skagit River, Washington, USA. *International Association of Sedimentologists Special Publication,* **30**, 285–298. https://doi.org/10.1002/9781444304251.ch14.

Ruga, M. R., Lanik, A., & Hults, C. P. (2020). *Focused Condition Assessment of Paleontological Resources within Lake Clark National Park and Preserve.* Fort Collins, CO: National Park Service. Natural Resource Report NPS/LACL/NRR—2020/2182.

Rumelhart, P. E., & Ingersoll, R. V. (1997). Provenance of the upper Miocene Modelo Formation and subsidence analysis of the Los Angeles Basin, southern California: Implications for paleotectonic and paleogeographic reconstructions. *Geological Society of America Bulletin,* **109**(7), 885–899. https://doi.org/10.1130/0016-7606(1997)109%3C0885:POTUMM%3E2.3.CO;2.

Salcido, C. J., Wilson, P., Tweet, J. S., et al. (2022). *Theodore Roosevelt National Park: Paleontological Resource Inventory (Public Version).* Fort Collins, CO: National Park Service. Natural Resource Report NPS/THRO/NRR—2022/2385. https://doi.org/10.36967/nrr-2293509.

Sandler, L. J., Chin, K., & Rogers. R. R. (1998). White River herbivore coprolites re-evaluated: Taphonomic and paleoecological implications. *Geological Society of America Abstracts with Programs,* **30**(7), A31.

Santucci, V. L. (1998). *The Yellowstone Paleontological Survey.* Yellowstone National Park, WY: Yellowstone Center for Resources. YCR-NR-91-1.

Santucci, V. L., Blodgett, R. B., Elder, W. P., Tweet, J. S., & Kenworthy, J. P. (2011). *Paleontological Resource Inventory and Monitoring: Central Alaska Network.* Fort Collins, CO: National Park Service. Natural Resource Technical Report NPS/NRSS/NRTR—2011/510.

Santucci, V. L., & Kenworthy, J. P. (2008). *Paleontological Resource Inventory and Monitoring: Southeast Alaska Network.* Fort Collins, CO: National Park Service.

Santucci, V. L., & Kenworthy, J. P. (2009). *Paleontological Resource Inventory and Monitoring: Klamath Network.* Fort Collins, CO: National Park Service. Natural Resource Technical Report NPS/NRPC/NRTR—2009/208.

Santucci, V. L., Kenworthy, J. P., & Visaggi, C. C. (2007). *Paleontological Resource Inventory and Monitoring, Chihuahuan Desert Network.* Denver, CO: National Park Service. TIC# D-500.

Santucci, V. L., Koch. A. L., & Kenworthy, J. (2004). Mojave National Preserve. In Santucci, V. L., Koch. A. L., & Kenworthy, J., eds.,

Paleontological Resource Inventory and Monitoring: Mojave Desert Network. Denver, CO: National Park Service, pp. 46–54. TIC# D-305.

Santucci, V. L., Tweet, J. S., Visaggi, C., & Hodnett, J.-P. M. (2024). The National Park System fossil record: Uncovering significant new paleontological discoveries through inventory, monitoring, research, and museum curation. *Parks Stewardship Forum*, **40**(1), 27–38. https://doi.org/10.5070/P540162919.

Santucci, V. L., & Wall, W. P. (1999). Paleontological resources at Grand Teton National Park, northwestern Wyoming. *University of Wyoming National Park Service Research Center Annual Report*, **22**, 31–36. https://doi.org/10.13001/uwnpsrc.1998.3357.

Schneider, M. C. (1967). Early Tertiary continental sediments of central and south-central Utah. *Brigham Young University Research Studies, Geology Series*, **14**, 143–194.

Scott, R. A. (1954). Fossil fruits and seeds from the Eocene Clarno Formation of Oregon. *Palaeontographica, Abteilung B*, **96**, 66–97.

Scott, T. W., Swift, D. J. P., Whittecar, G. R., & Brook, G. A. (2010). Glacio-isostatic influences on Virginia's late Pleistocene coastal plain deposits. *Geomorphology* **116**(1–2), 175–188. https://doi.org/10.1016/j.geomorph.2009.10.017.

Shattuck, G. B. (1904). Geological and paleontological relations, with a review of earlier investigations; the Miocene deposits of Maryland. In *Miocene*. Baltimore, MD: Johns Hopkins Press, pp. xxxiii–cxxxvii.

Shuler, J., & Santucci, V. L. (1986). Plant fossils discovered in Badlands National Park. *Park Science*, **6**(3), 6–7.

Singleton, S. W. (2008). Petrified wood in the Miocene Fleming Formation, Jasper County, Texas. *Gulf Coast Association of Geological Societies Transactions*, **58**, 797–814.

Smedes, H. W., & Prostka, H. J. (1972). Stratigraphic framework of the Absaroka Volcanic Supergroup in the Yellowstone National Park Region. *U.S. Geological Survey Professional Paper*, **729-C**, C1–C33. https://doi.org/10.3133/pp729C.

Smiley, C. J., & Rember, W. C. (1985). Composition of the Miocene Clarkia flora. In Smiley, C. J., ed., *Late Cenozoic History of the Pacific Northwest*. San Francisco, CA: Pacific Division, American Association for the Advancement of Science, pp. 95–112.

Smith, D. M. (2008). A comparison of plant-insect associations in the middle Eocene Green River Formation and the Upper Eocene Florissant Formation and their climatic implications. *Geological Society of America Special Paper*, **435**, 89–103.

Smith, M. A. (2023). *Climate and Diversity of Early and Middle Eocene Floras from Middle Latitudes of the Northern Hemisphere*. Gainesville: University of Florida, doctoral dissertation.

Smith, M. A., Greenwalt, D. E., & Manchester, S. R. (2023). Diverse fruits and seeds of the mid-Eocene Kishenehn Formation, northwestern Montana, USA, and their implications for biogeography. *Fossil Imprint*, **79**(1), 37–88. https://doi.org/10.37520/fi.2023.004.

Snow, J. K., & Lux, D. R. (1999). Tectono-sequence stratigraphy of Tertiary rocks in the Cottonwood Mountains and northern Death Valley area, California and Nevada. *Geological Society of America Special Paper*, **333**, 17–64. https://doi.org/10.1130/0-8137-2333-7.17.

Soper, E. K. (1938). Geology of the central Santa Monica mountains, Los Angeles County. *California Journal of Mines and Geology*, **34**(2), 131–180.

Spackman, W., Jr. (1951). Report on referred fossils [shipment MG-51-2]. USGS memo to D. M. Hopkins, February 19, 1951.

Spooner, A. M., Brubaker, L. B., & Foit Jr., F. F. (2008). *Ridley Lake: A Sediment Record of Holocene Vegetation and Climate History in North Cascades National Park Service Complex, Washington*. Report submitted to North Cascades National Park Service Complex in partial fulfillment of Task Agreement No. J9W88030002.

Spurr, J. E. (1903). Descriptive geology of Nevada south of the fortieth parallel and adjacent portions of California. *U.S. Geological Survey Bulletin*, **208**, 1–229. https://doi.org/10.3133/b208.

Staatz, M. H., Tabor, R. W., Weis, P. L., et al. (1972). Geology and mineral resources of the northern part of the North Cascades National Park, Washington. *U.S. Geological Survey Bulletin*, **1359**, 1–132. https://doi.org/10.3133/b1359.

Stadum, C. J., & Weigand, P. W. (1999). Fossil wood from the Middle Miocene Conejo Volcanics, Santa Monica Mountains, California. *Southern California Academy of Sciences Bulletin*, **98**, 15–25. https://doi.org/10.3160/0038-3872-98.1.15.

Stanton, R. J., & Alderson, J. M. (2013). Limestone interbedded with submarine volcanics: The Early–Middle Miocene Conejo Volcanics, California. *Facies*, **59**(3), 467–480. https://doi.org/10.1007/s10347-012-0320-6.

Stearns, H. T. (1928). Craters of the Moon National Monument, Idaho. *Idaho Geological Survey Bulletin*, **13**, 1–57.

Steere, W. C. (1947). Cenozoic and Mesozoic bryophytes of North America. *The American Midland Naturalist*, **36**, 298–324. https://doi.org/10.2307/2421507.

Stenzel, H. B., Krause, E. K., & Twining, J. T. (1957). *Pelecypoda from the Type Locality of the Stone City Beds (Middle Eocene) of Texas*. Austin: University of Texas Bureau of Economic Geology.

Stoneley, R. (1967). The structural development of the Gulf of Alaska sedimentary province in southern Alaska. *Quarterly Journal of the Geological Society*, **123**, 25–57. https://doi.org/10.1144/gsjgs.123.1.0025.

Strömberg, C. A. E. (2002). The origin and spread of grass-dominated ecosystems in the late Tertiary of North America: Preliminary results concerning the evolution of hypsodonty. *Palaeogeography, Palaeoclimatology, Palaeoecology*, **177**(1–2), 59–75. https://doi.org/10.1016/S0031-0182(01)00352-2.

Strömberg, C. A. E. (2004). Using phytolith assemblages to reconstruct the origin and spread of grass-dominated habitats in the Great Plains of North America during the Late Eocene to Early Miocene. *Palaeogeography, Palaeoclimatology, Palaeoecology*, **207**(3–4), 239–275. https://doi.org/10.1016/j.palaeo.2003.09.028.

Strömberg, C. A. E. (2011). Evolution of grasses and grassland ecosystems. *Annual Review of Earth and Planetary Sciences*, **39**, 517–544. https://doi.org/10.1146/annurev-earth-040809-152402.

Strömberg, C. A. E., & McInerney, F. A. (2011). The Neogene transition from C3 to C4 grasslands in North America: Assemblage analysis of fossil phytoliths. *Paleobiology*, **37**(1), 50–71. https://doi.org/10.1666/09067.1.

Strong, M. F. (1975). Little Fenner Valley. *Desert Magazine of the Southwest*, **38**(3), 8–11.

Stults, D. Z., & Axsmith, B. (2015). New plant fossil records and paleoclimate analyses of the late Pliocene Citronelle Formation flora, U.S. Gulf Coast. *Palaeontologia Electronica*, **18.3.47A**, 1–35. https://doi.org/10.26879/550.

Stults, D. Z., Tiffney, B. H., & Axsmith, B. J. (2022). New observations on the last *Pterocarya* (Juglandaceae) occurrences in eastern North America. *International Journal of Plant Sciences*, **183**(5), 380–392. https://doi.org/10.1086/720182.

Stults, D. Z., Wagner-Cremer, F., & Axsmith, B. J. (2011). Atmospheric paleo-CO_2 estimates based on *Taxodium distichum* (Cupressaceae) fossils from the Miocene and Pliocene of eastern North America. *Palaeogeography, Palaeoclimatology, Palaeoecology*, **309**, 327–332. https://doi.org/10.1016/j.palaeo.2011.06.017.

Swinehart, J. B., & Diffendal Jr., R. F. (1997). Geologic map of the Scottsbluff 1° x 2° quadrangle, Nebraska and Colorado. *U.S. Geological Survey Miscellaneous Investigations Series*, **2545**, 1 sheet, scale 1:250,000. https://doi.org/10.3133/i2545.

Tabor, R. W., Haugerud, R. A., Hildreth, W., & Brown, E. H. (2003). Geologic map of the Mount Baker 30- by 60-minute quadrangle, Washington. *U.S. Geological Survey Geologic Investigations Series Map*, **2660**, 1–70, 2 sheets, scale 1:100,000. https://doi.org/10.3133/i2660.

Tang, H., Liu, J., Wu, F. X., et al. (2019). Extinct genus *Lagokarpos* reveals a biogeographic connection between Tibet and other regions in the Northern Hemisphere during the Paleogene. *Journal of Systematics and Evolution*, **57** (6), 670–677. https://doi.org/10.1111/jse.12505.

Taylor, W. J. (1993). Stratigraphic and lithologic analysis of the Claron Formation in southwestern Utah. *Utah Geological Survey Miscellaneous Publication*, **93-1**, 1–52.

Tedford, R. H., Swinehart, J. B., Swisher III, C. C., et al. (1996). The Whitneyan-Arikareean transition in the High Plains. In Prothero, D. R., & Emry, R. J., eds., *The Terrestrial Eocene–Oligocene Transition in North America*. New York: Cambridge University Press, pp. 312–334.

Terry, D. O. Jr., & Evans, J. E. (1994). Pedogenesis and paleoclimatic implications of the Chamberlain Pass Formation, basal White River Group, Badlands of South Dakota. *Palaeogeography, Palaeoclimatology, Palaeoecology*, **110** (3–4), 197–215. https://doi.org/10.1016/0031-0182(94)90084-1.

Thomasson, J. R. (1980). *Palaoeriocoma* (Gramineae, Stipeae) from the Miocene of Nebraska: taxonomic and phylogenic significance. *Systematic Botany*, **5**(3), 233–240. https://doi.org/10.2307/2418370.

Thomasson, J. R. (1987). Late Miocene plants from northeastern Nebraska. *Journal of Paleontology*, **61**(5), 1065–1079.

Thornberry-Ehrlich, T. (2011). *Marsh-Billings-Rockefeller National Historical Park: Geologic Resources Inventory Report*. Fort Collins, CO: National Park Service. Natural Resource Report NPS/NRSS/GRD/NRR—2011/454.

Thornberry-Ehrlich, T. L. (2018). *Big Thicket National Preserve: Geologic Resources Inventory Report*. Fort Collins, CO: National Park Service. Natural Resource Report NPS/NRSS/GRD/NRR—2018/1779.

Tiffney, B. H. (1976). A survey of paleobotanical sites within the proposed Yukon–Charley National Rivers area. Center for Northern Studies, Wolcott, Vermont. In Young, S. B., ed., *The Environment of the Yukon–Charley Rivers Area, Alaska*. Wolcott, VT: Center for North Studies, Contributions from The Center for Northern Studies, **9**, pp. 173–203.

Tiffney, B. H. (1985). The Eocene North Atlantic Land Bridge: Its importance in Tertiary and modern phytogeography of the Northern Hemisphere. *Journal of the Arnold Arboretum* **66**(2), 243–273.

Tiffney, B. H. (1994). Re-evaluation of the age of the Brandon Lignite (Vermont, USA) based on plant megafossils. *Review of Palaeobotany and Palynology*, **82** (3–4), 299–315. https://doi.org/10.1016/0034-6667(94)90081-7.

Tiffney, B. H., & Manchester, S. R. (2001). The use of geological and paleontological evidence in evaluating plant phylogeographic hypotheses in the Northern Hemisphere Tertiary. *International Journal of Plant Sciences*, **162** (S6), S3–S17. https://doi.org/10.1086/323880.

Ting, F. T. (1972). Petrified peat from a Paleocene lignite in North Dakota. *Science*, **177**(4044), 165–166. https://doi.org/10.1126/science.177.4044.165.

Tomsich, C. S., McCarthy, P. J., Fowell, S. J., & Sunderlin, D. (2010). Paleofloristic and paleoenvironmental information from a Late Cretaceous (Maastrichtian) flora of the lower Cantwell Formation near Sable Mountain, Denali National Park, Alaska. *Palaeogeography, Palaeoclimatology, Palaeoecology*, **295**, 389–408. https://doi.org/10.1016/j.palaeo.2010.02.023.

Trexler, J. H., Jr. (1989). Ancestral Klamath River deposits at Gold Bluffs Prairie Creek Redwood State Park. *California Geology*, **42**, 147–152.

Troost, K. G. (2002). Summary of the Olympia nonglacial interval (MIS 3) in the Puget Lowland, Washington. *Geological Society of America Abstracts with Programs*, **34**(5), 109.

Tschudy, R. H., Pillmore, C. L., Orth, C. J., Gilmore, J. S., & Knight, J. D. (1984). Disruption of the terrestrial plant ecosystem at the Cretaceous-Tertiary Boundary, Western Interior. *Science*, **225**(4666), 1030–1032. https://doi.org/10.1126/science.225.4666.1030.

Tucker, S., & Voorhies, M. R. (2005). A diverse Late Miocene (Hemphillian) insectivore fauna from north-central Nebraska. *Journal of Vertebrate Paleontology*, **25**(supplement to 3), 124A–125A.

Turner, K. J., Berry, M. E., Page, W. R., et al. (2011). Geologic map of Big Bend National Park, Texas. *U.S. Geological Survey Scientific Investigations Map*, **3142**, 1–84, 1 sheet, scale 1:75,000. https://doi.org/10.3133/sim3142.

Tweet, J. S. (2016). *Trip Report: FOUS Technical Assistance Request #2214*. Internal report to Fort Union Trading Post National Monument.

Tweet, J. S., Flora, H., Weeks, S. R., McIntyre, E., & Santucci, V. L. (2021). *Grand Canyon-Parashant National Monument: Paleontological Resource Inventory (Public Version)*. Fort Collins, CO: National Park Service. Natural Resource Report NPS/PARA/NRR—2021/2338. https://doi.org/10.36967/nrr-2289972.

Tweet, J. S., & Santucci, V. L. (2017). *A Protocol for Paleontological Resource Monitoring at George Washington Birthplace National Monument*. Fort Collins, CO: National Park Service. Natural Resource Report NPS/GEWA/NRR—2017/1526.

Tweet, J. S., Santucci, V. L., & Connors, T. (2012b). *Paleontological Resource Inventory and Monitoring: Mediterranean Coast Network*. Fort Collins, CO: National Park Service. Natural Resource Technical Report NPS/MEDN/NRTR—2012/640.

Tweet, J. S., Santucci, V. L., & Connors, T. (2013). *Paleontological Resource Inventory and Monitoring: Greater Yellowstone Network*. Fort Collins, CO: National Park Service. Natural Resource Technical Report NPS/GRYN/NRTR—2013/794.

Tweet, J. S., Santucci, V. L., & Connors, T. (2014). *Paleontological Resource Inventory and Monitoring: Northeast Coastal and Barrier Network*. Fort Collins, CO: National Park Service. Natural Resource Technical Report NPS/NCBD/NRTR— 2014/897.

Tweet, J. S., Santucci, V. L., & Connors, T. (2015). *Paleontological Resource Inventory and Monitoring: Southern Plains Network*. Fort Collins, CO: National Park Service. Natural Resource Technical Report NPS/SOPN/NRR—2015/971.

Tweet, J. S., Santucci, V. L., & Connors, T. (2016). *Paleontological Resource Inventory and Monitoring: Mojave Desert Network*. Fort Collins, CO: National Park Service. Natural Resource Report NPS/MOJN/NRR—2016/1209.

Tweet, J. S., Santucci, V. L., Connors, T., & Kenworthy, J. P. (2012c). *Paleontological Resource Inventory and Monitoring: Northern Colorado Plateau Network*. Fort Collins, CO: National Park Service. Natural Resource Technical Report NPS/NCPN/NRTR—2012/585.

Tweet, J. S., Santucci, V. L., Convery, K., Hoffman, J., & Kirn, L. (2020). *Channel Islands National Park: Paleontological Resource Inventory (Public Version)*. Fort Collins, CO: National Park Service. Natural Resource Report NPS/CHIS/NRR—2020/2171. https://doi.org/10.36967/nrr-2278664.

Tweet, J. S., Santucci, V. L., & Hunt, A. P. (2012a). An inventory of packrat (*Neotoma* spp.) middens in National Park Service areas. *New Mexico Museum of Natural History and Science Bulletin*, **57**, 355–368.

Tweet, J. S., Santucci, V. L., & Kenworthy, J. P. (2008). *Paleontological Resource Inventory and Monitoring: Sonoran Desert Network*. Fort Collins, CO: National Park Service. Natural Resource Technical Report NPS/NRPC/NRTR—2008/130.

Tweet, J. S., Santucci, V. L., & Kenworthy, J. P. (2010). *Paleontological Resource Inventory and Monitoring: Northeast Temperate Network*. Fort Collins, CO: National Park Service. Natural Resource Technical Report NPS/NRPC/NRTR—2010/326.

Tweet, J. S., Santucci, V. L., & Kenworthy, J. P. (2011). *Paleontological Resource Inventory and Monitoring: Northern Great Plains Network*. Fort Collins, CO: National Park Service. Natural Resource Technical Report NPS/NRPC/NRTR—2011/437.

Tweet, J. S., Santucci, V. L., Kenworthy, J. P., & Mims, A. L. (2009). *Paleontological Resource Inventory and Monitoring: Southern Colorado Plateau Network*. Fort Collins, CO: National Park Service. Natural Resource Technical Report NPS/NRPC/NRTR—2009/245.

Uchupi, E., Giese, G. S., Aubery, D. G., & Kim, D. J. (1996). The late Quaternary construction of Cape Cod, Massachusetts: A reconsideration of the W.M. Davis model. *Geological Society of America Special Paper*, **309**, 1–69. https://doi.org/10.1130/0-8137-2309-4.1.

Vandiver, V. W. (1936). Geological report on White Sands National Monument. *Southwestern Monuments Special Report*, **5**, 381–400.

ver Straeten, C. A., Monteverde, D. H., & Inners, J. D., leaders. (2001). Road log and stop descriptions; Day 2, Bushkill access area: The power of water – a brief history of the Delaware Water Gap National Recreation Area. In Inners, J. D., & Fleeger, G. M., eds., *2001: A Delaware River Odyssey*. Harrisburg: Field Conference of Pennsylvania Geologists, pp. 250–253.

Vogt, P. R., & Eshelman, R. E. (1987). Maryland's Cliffs of Calvert: A fossiliferous record of mid-Miocene inner shelf and coastal environments. *Geological Society of America Centennial Field Guide*, **5**, 9–14. https://doi.org/10.1130/0-8137-5405-4.9.

Voorhies, M. R. (1990). *Vertebrate Paleontology of the Proposed Norden Reservoir Area, Brown, Cherry, and Keya Paha Counties, Nebraska*. Lincoln: Division of Archeological Research, University of Nebraska. Technical Report 82-09.

Voorhies, M. R., & Corner. R. G. (1993). An inventory and evaluation of vertebrate paleontological sites along the Niobrara/Missouri scenic riverways project corridors. Unpublished report to National Park Service.

Voorhies, M. R., & Thomasson, J. R. (1979). Fossil grass anthoecia within Miocene rhinoceros skeletons: Diet in an extinct species. *Science*, **206**(4416), 331–333. https://doi.org/10.1126/science.206.4416.33.

Wahrhaftig, C., Wolfe, J. A., Leopold, E. B., & Lanphere, M. A. (1969). The coal-bearing group of the Nenana Coal Field, Alaska. *U.S. Geological Survey Bulletin*, **1274-D**, D1–D30. https://doi.org/10.3133/b1274D.

Wang, Y., & Manchester, S. R. (2000). *Chaneya*, a new genus of winged fruit from the Tertiary of North America and eastern Asia. *International Journal of Plant Sciences*, **161**(1), 167–178. https://doi.org/10.1086/314227.

Watson, G. E. (2006). *Big Thicket Plant Ecology: An Introduction*. Denton, TX: Big Thicket Association.

Weaver, C. E. (1949). Geology and mineral deposits of an area north of San Francisco Bay. *California Division of Mines and Geology Bulletin*, **149**, 1–135.

Westerhold, T., Marwan, N., Drury, A. J., et al. (2020). An astronomically dated record of Earth's climate and its predictability over the last 66 million years. *Science*, **369**(6509), 1383–1387. https://doi.org/10.1126/science.aba6853.

Wetterich, S., Grosse, G., Schirrmeister, L., et al. (2012). Late Quaternary environmental and landscape dynamics revealed by a pingo sequence on the northern Seward Peninsula, Alaska. *Quaternary Science Reviews*, **39**, 26–44. https://doi.org/10.1016/j.quascirev.2012.01.027.

Wheeler, E. A. (1991). Paleocene dicotyledonous trees from Big Bend National Park, Texas: variability in wood types common in the Late Cretaceous and Early Tertiary, and ecological inferences. *American Journal of Botany*, **78**(5), 658–671. https://doi.org/10.2307/2445087.

Wheeler, E. A. (2001). Fossil dicotyledonous woods from Florissant Fossil Beds National Monument, Colorado. *Proceedings of the Denver Museum of Natural History, Series 4*, **1**, 187–203.

Wheeler, E. A. (2024). Fossil woods of Yellowstone National Park. *Parks Stewardship Forum*, **40**(1), 181–195. https://doi.org/10.5070/P540162929.

Wheeler, E. A., & Lehman, T. M. (2005). Upper Cretaceous–Paleocene conifer woods from Big Bend National Park, Texas. *Palaeogeography, Palaeoclimatology, Palaeoecology*, **226**(3), 233–258. https://doi.org/10.1016/j.palaeo.2005.05.014.

Wheeler, E. A., & Lehman, T. M. (2009). New Late Cretaceous and Paleocene dicot woods of Big Bend National Park, Texas and review of Cretaceous wood characteristics. *IAWA Journal*, **30**(3), 293–318. https://doi.org/10.1163/22941932-90000220.

Wheeler, E. A., & Manchester, S. R. (2002). Woods of the middle Eocene nut beds flora, Clarno Formation, Oregon, USA. *IAWA Journal*, **Supplement 3**, 1–188.

Wheeler, E. A., & Manchester, S. R. (2014). Middle Eocene trees of the Clarno Petrified Forest, John Day Fossil Beds National Monument, Oregon. *Paleobios*, **30**(3). https://doi.org/10.5070/P9303022471.

Wheeler, E. A., & Meyer, H. W. (2012). A new (*Hovenia*) and an old (*Chadronoxylon*) fossil wood from the Late Eocene Florissant Formation, Colorado, U.S.A. *IAWA Journal*, **33**(3), 309–318. https://doi.org/10.1163/22941932-90000096.

Wheeler, E. F., Scott, R. A., & Barghoorn, E. S. (1977). Fossil dicotyledonous woods from Yellowstone National Park. *Journal of the Arnold Arboretum*, **58**(3), 280–306.

Wheeler, E. F., Scott, R. A., & Barghoorn, E. S. (1978). Fossil dicotyledonous woods from Yellowstone National Park, II. *Journal of the Arnold Arboretum*, **59**(1), 1–31.

Wick, S. L. (2021). Paleontological inventory of Paleozoic, Late Mesozoic, and Cenozoic plant, invertebrate, and vertebrate fossil species from Big Bend National Park, Texas, USA: Over a century of paleontological discovery. *Zitteliana*, **95**, 95–134. https://doi.org/10.3897/zitteliana.95.73026.

Wick, S. L., & Corrick, D. W. (2015). *Paleontological Inventory of Big Bend National Park, Texas: The Place, the People, and the Fossils*. Internal report to Big Bend National Park Division of Science and Resource Management.

Wiemann, M. C., Manchester, S. R., & Wheeler, E. A. (1999). Paleotemperature estimation from dicotyledonous wood anatomical characters. *Palaios*, **14**(5), 459–474. https://doi.org/10.2307/3515397.

Wilf, P. (2000). Late Paleocene–Early Eocene climate changes in southwestern Wyoming: paleobotanical analysis. *Geological Society of America Bulletin*, **112**(2), 292–307. https://doi.org/10.1130/0016-7606(2000)112<292:LPECCI>2.0.CO;2.

Wilf, P., Carvalho, M. R., & Stiles, E. (2023). The end-Cretaceous plant extinction: Heterogeneity, ecosystem transformation, and insights for the future. *Cambridge Prisms: Extinction*, **1**, e14. https://doi.org/10.1017/ext.2023.13.

Wilf, P., & Johnson, K. R. (2004). Land plant extinction at the end of the Cretaceous: A quantitative analysis of the North Dakota megafloral record. *Paleobiology*, **30**(3), 347–368. https://doi.org/10.1666/0094-8373(2004)030<0347:LPEATE>2.0.CO;2.

Wilson, G. A. (1988). *The Effects of Subsurface Dissolution of Permian Salt on the Deposition, Stratigraphy and Structure of the Ogallala Formation (Late Miocene Age), Northeast Potter County, Texas*. Canyon: West Texas State University, master's thesis.

Wilson, F. H., Dover, J. H., Bradley, D. C., et al. (1998). Geologic map of central (interior) Alaska. *U.S. Geological Survey Open-File Report*, **98-133-A**, 1–64. https://doi.org/10.3133/ofr98133A.

Wilson, J. A., & Runkel, A. C. (1989). *Field Guide to Paleocene Stratigraphy and Vertebrate Paleontology of the Big Bend Region*. Internal report to the University of Texas at Austin.

Wilson, L. R. (1932). The Two Creeks Forest Bed, Manitowoc County, Wisconsin. *Transactions of the Wisconsin Academy of Sciences, Arts, and Letters*, **27**, 31–46. https://doi.org/10.2307/2481158.

Wilson, L. R. (1936). Further fossil studies of the Two Creeks Forest Bed, Manitowoc County, Wisconsin. *Bulletin of the Torrey Botanical Club*, **63**(6), 317–325. https://doi.org/10.2307/2481158.

Wing, S. L. (1987). Eocene and Oligocene floras and vegetation of the Rocky Mountains. *Annals of the Missouri Botanical Garden*, **74**(4), 748–784. https://doi.org/10.2307/2399449.

Wing, S. L. (1998). Tertiary vegetation of North America as a context for mammalian evolution. In Janis, C. M., Scott, K. M., & Jacobs, L. L., eds., *Evolution of Tertiary Mammals of North America, Volume 1: Terrestrial Carnivores, Ungulates, and Ungulatelike Mammals*. Cambridge: Cambridge University Press, pp. 37–60.

Wing, S. L., Alroy, J., & Hickey, L. J. (1995). Plant and mammal diversity in the Paleocene to Early Eocene of the Bighorn Basin. *Palaeogeography, Palaeoclimatology, Palaeoecology*, **115**(1–4), 117–155. https://doi.org/10.1016/0031-0182(94)00109-L.

Wing, S. L., & Boucher, L. D. (1998). Ecological aspects of the Cretaceous flowering plant radiation. *Annual Review of Earth and Planetary Sciences*, **26**(1), 379–421. https://doi.org/10.1146/annurev.earth.26.1.379.

Wing, S. L., & Currano, E. D. (2013). Plant response to a global greenhouse event 56 million years ago. *American Journal of Botany*, **100**(7), 1234–1254. https://doi.org/10.3732/ajb.1200554.

Wing, S. L., Harrington, G. J., Smith, F. A., et al. (2005). Transient floral change and rapid global warming at the Paleocene-Eocene Boundary. *Science*, **310**(5750), 993–996. https://doi.org/10.1126/science.1116913.

Wing, S. L., & Hickey, L. J. (1984). The *Platycarya* perplex and the evolution of the Juglandaceae. *American Journal of Botany*, **71**(3), 388–411. https://doi.org/10.1002/j.1537-2197.1984.tb12525.x.

Wingate, F. H., & Nichols, D. J. (2001). Palynology of the uppermost Eocene lacustrine deposits at Florissant Fossil Beds National Monument, Colorado. *Proceedings of the Denver Museum of Nature and Science, Series 4*, **1**, 71–135.

Winkler, M. G. (1985). A 12,000 year history of vegetation and climate for Cape Cod, Massachusetts. *Quaternary Research*, **23**(3), 301–312. https://doi.org/10.1016/0033-5894(85)90037-7.

Wolfe, J. A. (1964). Sample No. 6319 of Whistler and Wolfe 7/19/63 – USGS No. Paleobot. loc. 9935 [11/16/1964]. Alaska Paleontological Database. www.alaskafossil.org/.

Wolfe, J. A. (1966). Tertiary plants from the Cook Inlet Region, Alaska. *U.S. Geological Survey Professional Paper*, **398-B**, B1–B32. https://doi.org/10.3133/pp398B.

Wolfe, J. A. (1968). Sample No. 67AR259 – USGS No. Paleobot. Loc. 11125 [05/03/1968]. Alaska Paleontological Database. www.alaskafossil.org/.

Wolfe, J. A. (1970). Sample No. 68APr 130 C – USGS No. Paleobot. loc. 11183 [02/16/1970]. Alaska Paleontological Database. www.alaskafossil.org/.

Wolfe, J. A. (1972). An interpretation of Alaskan Tertiary floras. In Graham, A. ed., *Floristics and Paleofloristics of Asia and Eastern North America*. Amsterdam: Elsevier Publishing Co., pp. 201–233.

Wolfe, J. A. (1976). Report on referred fossils. Sample No. 75-Apr-37 [06/24/1976]. Alaska Paleontological Database. www.alaskafossil.org/.

Wolfe, J. A. (1977). Paleogene floras from the Gulf of Alaska Region. *U.S. Geological Survey Professional Paper*, **997**, 1–108. https://doi.org/10.3133/pp997.

Wolfe, J. A. (1980). Report on referred fossils [Tolstoi Formation, Shipments A-77-39M, A-80-20M, A-80-21M, 07/15/1980, ID 607]. Alaska Paleontological Database. www.alaskafossil.org/.

Wolfe, J. A. (1981). A chronologic framework for Cenozoic megafossil flora of northwestern North America and its relation to marine geochronology. *Geological Society of America Special Paper*, **194**, 39–47. https://doi.org/10.1130/SPE184-p39.

Wolfe, J. A. (1992). Climatic, floristic, and vegetational changes near the Eocene/Oligocene Boundary in North America. In Prothero, D. R., & Berggren, W. A., eds., *Eocene-Oligocene Climatic and Biotic Evolution*. Princeton, NJ: Princeton University Press, pp. 421–436. https://doi.org/10.1515/9781400862924.421.

Wolfe, J. A. (1994a). Alaskan Palaeogene climates as inferred from the CLAMP database. In Boulter, M. C., & Fisher, H. C., eds., *Cenozoic Plants and Climates of the Arctic*. Berlin: Springer-Verlag, NATO ASI Series, **127**, pp. 223–237.

Wolfe, J. A. (1994b). An analysis of Neogene climates in Beringia. *Paleogeography, Paleoclimatology, Paleoecology*, **108**, 207–216. https://doi.org/10.1016/0031-0182(94)90234-8.

Wolfe, J. A., Hopkins, D. M., & Leopold, E. B. (1966). Tertiary stratigraphy and paleobotany of the Cook Inlet Region, Alaska. *U.S. Geological Survey Professional Paper*, **398-A**, A1–A29. https://doi.org/10.3133/pp398A.

Wolfe, J. A., & Tanai, T. (1980). The Miocene Seldovia Point flora from the Kenai Group, Alaska. *U.S. Geological Survey Professional Paper*, **1105**, 1–52. https://doi.org/10.3133/pp1105.

Wolfe, J. A., & Tanai, T. (1987). Systematics, phylogeny, and distribution of *Acer* (maples) in the Cenozoic of western North America. *Journal of the Faculty of Science, Hokkaido University. Series 4, Geology and Mineralogy*, **22**, 1–246.

Wolfe, J. A., & Upchurch, G. R. (1986). Vegetation, climatic and floral changes at the Cretaceous-Tertiary boundary. *Nature*, **324**(6093), 148–152. https://doi.org/10.1038/324148a0.

Wolfe, J. A., & Upchurch, G. R. (1987). Leaf assemblages across the Cretaceous-Tertiary boundary in the Raton Basin, New Mexico and Colorado. *Proceedings of the National Academy of Sciences*, **84**(15), 5096–5100. https://doi.org/10.1073/pnas.84.15.5096.

Wolfe, J. A., & Wehr, W. (1988). Rosaceous *Chamaebatiaria*-like foliage from the Paleogene of western North America. *Aliso: A Journal of Systematic and Floristic Botany*, **12**(1), 177–200. https://doi.org/10.5642/aliso.19881201.14.

Woodburne, M. O. (2007). Phyletic diversification of the *Cormohipparion occidentale* complex (Mammalia, Perissodactyla, Equidae), Late Miocene, North America, and the origin of the Old World *Hippotherium* datum. *Bulletin of the American Museum of Natural History*, **306**, 1–138. https://doi.org/10.1206/0003-0090(2007)306[1:PDOTCO]2.0.CO;2.

Woodcock, D., & Kalodimos, N. (2003). The tree molds at Puuhonua o Honaunau: A record of loulu palms on the Kona coast. *Park Paleontology*, **7**(1), 2–4.

Woodcock, D., & Kalodimos, N. (2005). Tree mold evidence of Loulu palm (*Ptitchardia* sp.) forest on the Kona Coast, Hawai'i. *Pacific Science*, **59**(4), 491–498. https://doi.org/10.1353/psc.2005.0053.

Wrenn, J. H., & DiBenedetto, J. N. (2002). Palynology of the Scenic Member, Brule Formation, Sage Creek Wilderness Area, Badlands National Park, Interior, South Dakota. *Palynology*, **26**, 280.

Yerkes, R. F., & Campbell, R. H. (1979). Stratigraphic nomenclature of the central Santa Monica Mountains, Los Angeles County, California. *U.S. Geological Survey Bulletin*, **1457**, E1–E31. https://doi.org/10.3133/b1457E.

Yuretich, R. F. (1984). Yellowstone fossil forests: New evidence for burial in place. *Geology*, **12**(3), 159–162. https://doi.org/10.1130/0091-7613(1984)12%3C159:YFFNEF%3E2.0.CO;2.

Zachos, J. C., Dickens, G. R., & Zeebe, R. E. (2008). An Early Cenozoic perspective on greenhouse warming and carbon-cycle dynamics. *Nature*, **451**(7176), 279–283. https://doi.org/10.1038/nature06588.

Zeigler, J. M., Tuttle, S. D., Tasha, H. J., & Giese. G. S. (1964). Pleistocene geology of outer Cape Cod, Massachusetts. *Geological Society of America*

Bulletin, **75**(8), 705–714. https://doi.org/10.1130/0016-7606(1964)75[705:PGOOCC]2.0.CO;2.

Zetter, R., Farabee, M. J., Pigg, K. B., et al. (2011). Palynoflora of the late Paleocene silicified shale at Almont, North Dakota, USA. *Palynology*, **35**(2), 179–211. https://doi.org/10.1080/gspalynol.35.2.179.

Zuntini, A. R., Carruthers, T., Maurin, O., et al. (2024). Phylogenomics and the rise of the angiosperms. *Nature*, **629**(8013), 843–850. https://doi.org/10.1038/s41586-024-07324-0.

Acknowledgments

This inventory of paleobotanical resources from National Park Service areas would not have been possible without the previous research undertaken and articles published by scores of paleontologists, all of whom we have identified and cited in this publication. This project was supported through the National Park Service–Paleontological Society partnership known as the Paleontology in the Parks Fellowship Program. We would like to specifically recognize the leadership within the Paleontological Society for their support of this project and the fellowship, including Rowan Lockwood, College of William & Mary and past president of the Paleontological Society, and Brenda Hunda, Curator of Invertebrate Paleontology, Cincinnati Museum Center, Geier Collections & Research Center. We also appreciate the support of National Park Service staff within the Geologic Resources Division and those working at various parks and monuments. Finally, we extend our thanks to Dale Greenwalt and MacKenzie Smith for providing photos of fossils from Glacier National Park used in this publication.

Cambridge Elements

Elements of Paleontology

Series Editor-in-Chief

Brenda R. Hunda
Cincinnati Museum Center

About the Series

The Elements of Paleontology series is a publishing collaboration between the Paleontological Society and Cambridge University Press. The series covers the full spectrum of topics in paleontology and paleobiology, and related topics in the Earth and life sciences of interest to students and researchers of paleontology.

The Paleontological Society is an international nonprofit organization devoted exclusively to the science of paleontology: invertebrate and vertebrate paleontology, micropaleontology, and paleobotany. The Society's mission is to advance the study of the fossil record through scientific research, education, and advocacy. Its vision is to be a leading global advocate for understanding life's history and evolution. The Society has several membership categories, including regular, amateur/avocational, student, and retired. Members, representing some 40 countries, include professional paleontologists, academicians, science editors, Earth science teachers, museum specialists, undergraduate and graduate students, postdoctoral scholars, and amateur/avocational paleontologists.

Cambridge Elements

Elements of Paleontology

Elements in the Series

Testing Character Evolution Models in Phylogenetic Paleobiology: A Case Study with Cambrian Echinoderms
April Wright, Peter J. Wagner and David F. Wright

The Taphonomy of Echinoids: Skeletal Morphologies, Environmental Factors and Preservation Pathways
James H. Nebelsick and Andrea Mancosu

Virtual Paleontology: Tomographic Techniques For Studying Fossil Echinoderms
Jennifer E. Bauer, Imran A. Rahman

Follow The Fossils: Developing Metrics For Instagram As A Natural Science Communication Tool
Samantha B. Ocon, Lisa Lundgren, Richard T. Bex II, Jennifer E. Bauer, Mary Janes Hughes, and Sadie M. Mills

Niche Evolution and Phylogenetic Community Paleoecology of Late Ordovician Crinoids
Selina R. Cole and David F. Wright

Molecular Paleobiology of the Echinoderm Skeleton
Jeffrey R. Thompson

A Review of Blastozoan Echinoderm Respiratory Structures
Sarah L. Sheffield, Maggie R. Limbeck, Jennifer E. Bauer, Stephen A. Hill, and Martina Nohejlová

A Review and Evaluation of Homology Hypotheses in Echinoderm Paleobiology
Colin D. Sumrall, Sarah L. Sheffield, Jennifer E. Bauer, Jeffrey R. Thompson, and Johnny A. Waters

The Ecology of Biotic Interactions in Echinoids: Modern Insights into Ancient Interactions
Elizabeth Petsios, Lyndsey Farrar, Shamindri Tennakoon, Fatemah Jamal, Roger W. Portell, Michał Kowalewski, and Carrie L. Tyler

What Does Graptolite Origination and Extinction Reveal about the Cause of the Late Ordovician Mass Extinction?
Charles E. Mitchell, H. David Sheets, Michael J. Melchin and Chris Holmden

Critical History for Tomorrow's Paleontology
Victor Monnin

Cenozoic Paleobotanical Resource Inventory of the National Park System
T. P. Matel, I. B. Huegele, C. R. Cace, K. M. M. Bober, L. D. Boucher, V. E. McCoy, E. J. Hermsen, mS. R. Manchester, C. C. Visaggi, J. S. Tweet and V. L. Santucci

A full series listing is available at: www.cambridge.org/EPLY

For EU product safety concerns, contact us at Calle de José Abascal, 56–1°,
28003 Madrid, Spain or eugpsr@cambridge.org.

www.ingramcontent.com/pod-product-compliance
Ingram Content Group UK Ltd.
Pitfield, Milton Keynes, MK11 3LW, UK
UKHW021856010326
468546UK00018B/750